TIME'S ARROW AND EVOLUTION

TIME'S ARROW AND EVOLUTION

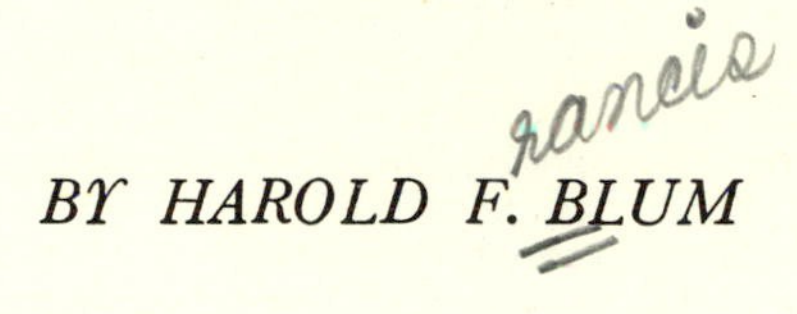

BY HAROLD F. BLUM

PRINCETON UNIVERSITY PRESS
PRINCETON, NEW JERSEY

Library of Congress Catalog Card: 68-31676
ISBN 0-691-02354-9 (paperback edn.)
ISBN 0-691-08061-5 (hardcover edn.)

Third edition, 1968

The expanded 1968 edition includes text revisions, a new Preface, additional entries in the Bibliography, and a new concluding chapter.

First Princeton Paperback Edition, 1968
Second Printing, 1970

PRINTED IN THE UNITED STATES OF AMERICA

To Mabel and Jannet

PREFACE

I FIRST began to think about possible relationships between the second law of thermodynamics and organic evolution during the summer of 1933 while a guest of the laboratory of the Collège de France, at Concarneau in Brittany. The previous spring I had discussed Henderson's *Fitness of the Environment* with my class in General Physiology at the University of California, and no doubt the seed was sown then in my mind. Shortly after my return home I presented some of my ideas at a meeting of geneticists. There was nothing to suggest that the audience found the concepts too radical, though perhaps the lack of response indicated that they were not worth the trouble of refuting. My paper was, however, duly reported by the press, and did cause a little stir. Numerous letters to the editor of one of the local papers acclaimed my having refuted Darwin. The letters became so numerous that the editor undertook to answer them, explaining—more clearly than I had—what I was really talking about, pointing out that nothing I had said denied evolution by natural selection. This incident has warned me against the possible misinterpretation of this book, and so I feel I should point out that, so far as I am aware, none of the ideas presented are in conflict, indeed they seem complementary to, the concepts of modern Darwinism.

Subsequently, the book made numerous false starts, and has suffered various delays. A great war has occurred, and there have been other vagaries of fortune to postpone its completion. In the intervening years many ideas have changed radically, particularly as regards cosmic evolution, so that much that was written only a few years ago does not hold today. This shift in point of view has emphasized the importance of putting the argument in a general form rather than tying it too securely to any particular theory. There has been nothing in all this change, however, to weaken the status of the second law of thermodynamics; rather its general applicability to the evolution of the nonliving world—and to that of the living world as well—has become more obvious.

Others have dealt with the relationship of the second law to organic evolution, both before and since my first essay into the field. At present, interest in the subject seems to be increasing, and many are concerned with the problem who have not put their thoughts in print.

This seems natural enough; for it is difficult to believe that the evolution of living things, a process that has proceeded unidirectionally in time, could be independent of the great principle of irreversibility. I have attempted in this book to examine various relationships between the second law of thermodynamics and organic evolution, and in so doing I may not in all cases have paid strict attention to the origin of the ideas involved. Hence it may seem that due concern has not been given to the writings of others, but by this time it would be impossible to untangle all the ideas and trace out their derivations. The least I can say is that I feel immeasurably in debt to those who have helped me to write this book. Many of these I have not known personally, our only contact being through what they have published. I have tried in my bibliography, at the end of the book, to include those references that have contributed importantly to my thinking in this regard, and also some that may assist the reader who wishes to explore farther into particular aspects. More specific citations have also been made in the footnotes.

A number of persons undertook to read the manuscript, or parts of it, in various stages of completion; and to offer criticisms which have been most valuable to me in making revision. It is a pleasure to acknowledge this kindness on the part of George W. Bain, John Tyler Bonner, A. F. Buddington, Marcel Florkin, Otto Glaser, David R. Goddard, Warren K. Green, H. H. Hess, Walter Kauzmann, Gordon M. Loos, Arthur K. Parpart, Newton L. Pierce, Harold H. Plough, and C. B. van Niel. My thanks are also due to Mrs. Dorothy D. Hollmann for her critical editorship of the manuscript, and to Mrs. Margie R. Matthews for preparation of drawings.

The book's final coming into being was made possible by a John Simon Guggenheim Memorial Foundation Fellowship which was granted for the purpose at the end of the war. And I must thank Mr. Henry Allen Moe for his kindly advice that I undertake the thing I really wanted most to do, rather than follow some more "practical" plan. Without this timely encouragement the work might have been postponed indefinitely.

I take this opportunity to thank my numerous hosts during the course of my Fellowship; the Departments of Physiology and Zoology at the University of California, the Department of Zoology of the University of California at Los Angeles, the Department of Biology at the University of Rochester, and the Marine Biological Laboratory at Woods Hole. Seminars which I gave before certain groups were helpful to me in crystalizing my ideas, and I am grateful for the comments I received at those times. The San Francisco Bay group known as the Biosystematists was one of these; the biology seminars at the

University of Rochester and at Harvard University were the others. I have completed the book while a member of the National Cancer Institute and of the Department of Biology of Princeton University—to both my grateful appreciation for their assistance and many courtesies. And I cannot end without again mentioning the place where it all began, among pleasant and stimulating surroundings at the Laboratoire de Zoologie Maritime, of the Collège de France, at Concarneau.

HAROLD F. BLUM

Woods Hole,
August 1950

PREFACE TO SECOND EDITION

IN MAKING the limited revision to which I am restricted at this time, I have the feeling that I am not doing full justice to the readers and reviewers who received the book so graciously when it appeared. They have been, on the whole, so kind in their comments that I do not like to let them down. A good many things have been learned in the meantime, and points of view have shifted in several fields, so that there is much new that could well be incorporated; but I have had to limit myself to those things that bear directly on the main argument. It has been borne home to me that in another three years many more revisions will be needed, for science moves rapidly these days; and this has, perhaps, kept me from taking as definite a stand as I might have in some places. But I have contented myself with the thought that in spite of the changes I have had to make, and the many more I could have made; the principal theme seems to stand without more than minor modification. The need to take the second law of thermodynamics into our thinking about evolution appears to me all the more certain; although I am less sure that I have said so as clearly as I could wish.

In presenting the argument in the first place, it seemed necessary that the reader should have a clear understanding, on at least an elementary level, of numerous aspects of the problem; and my attempt to provide this may have confused some readers as to my intention. It was perhaps unavoidable that a certain number of them should have found parts of the book too elementary, and other parts somewhat abstruse; since any specialist must find the handling of his own subject incomplete, while other fields, with which he is less familiar, may appear to him to be treated at a more advanced level.

That a good many readers have found something new in the book in the way of general approach is heartening; and if some have found only an assembly of more or less general information, I think I should not complain. The latter may have reason on their side, but I am encouraged to remember that on first reading *The Fitness of the Environment* there seemed to me very little in it that was new. But even to make so slight a comparison of the two books is a form of self-flattery in which I should not, perhaps, indulge.

The greatest amount of revision has been called for in the parts dealing with the origin of the earth, and the origin of life. In the latter case, I have added a fairly extensive addendum to the chapter concerned. I should have liked to say more; but have decided to leave the reader ample room for his own speculations, after cautioning him to stay within the bounds of physical principles, including the restrictions of the second law, and of the geochemical and biological evidence. In order to avoid rearrangement of the index, this addendum has been inserted without changing the original pagination. I have also added a number of recent items to the bibliography; reference is made to certain of these in the revised text, by the authors name and the date.

I am happy of this opportunity to thank the Princeton University Press, and particularly Mr. Herbert S. Bailey, Jr., for the fine job of manufacture of the book, and for many kindnesses.

HAROLD F. BLUM

Princeton
December 1954

PREFACE TO THE THIRD EDITION

The years since this book was first written have been packed with more scientific activity than any comparable period in the world's history, and many additions and revisions would have to be made to bring this edition completely "up to date." Many of the pertinent developments—particularly in biochemistry dealing with intermediate metabolic pathways and with the genetic code—have been treated so thoroughly elsewhere that I have assumed them to be available to all readers in one form or another. Therefore I have tried to make only such changes as might make the volume compatible with this later knowledge, and have not attempted extensive review.

What has concerned me in revising is the increasing trend toward "deterministic" thinking that has come with our scientific successes, often bringing with it tacit teleology and panglossianism. Implications of Information Theory that might give a more "probabilitic" view have had less influence. For example, analogies between computer operation and natural selection seem to have been generally neglected. I have, in an added chapter, pointed out some of these analogies, their bearing on the explanation of evolutionary processes, and the dangers of carrying analogy too far. Some of the things said in that chapter may seem to be repetitions from earlier ones, but I believe they can stand reemphasis.

A good many of the notions in the added chapter have been tossed about in pleasant conversations with Roger Pinkham, and I am not sure just what part is his and what mine, nor how much he would wish to take responsibility for. But I am happy to acknowledge my debt to him, and thank him for his criticism of the new chapter. I have also to thank Walter Kauzmann, Eugene McLaren, and Larry Mason for valuable criticisms of that chapter.

HAROLD F. BLUM

State University of New York
at Albany
March 1968

CONTENTS

TIME'S ARROW AND EVOLUTION

"*Li tens qui s'en va nuit et jor,*
Sans repos prendre et sans sejor—"
—WILLIAM OF LORRIS

I · PERSPECTIVES

"*Ah, but a man's reach should exceed his grasp.*"—ROBERT BROWNING

THE idea of evolution from a past primitive state to a present more complex one brings unity to biology, explaining the relationships among living organisms and the recurrent patterns one finds throughout the living world whether it be explored with binocular, test-tube, microscope, or Geiger-Müller counter. The concept is not unique to biology; the astronomer regards the universe in the same way, and the geologist, and others too, find evolution essential to the understanding of the broader aspects of their subjects.

The beginning of widespread influence of the evolutionary concept in biology is associated with the middle of the nineteenth century and the name of Charles Darwin. Earlier in that century the general idea of evolution was not uncommon in natural science, but an acceptable mechanism for explaining organic evolution had been wanting. The introduction of the idea of variation and natural selection by Darwin provided a rationalization, serving to crystallize what before had been a vague recognition of some sort of historical trend. The original Darwinian ideas underwent various modifications, but have survived as a part of our modern concept of evolution which has been given a firm basis by the science of genetics and the introduction of statistical treatment of populations. There are few biologists today—and the present author is certainly not one—who would venture to attack this solid structure.

One may ask, however, if all the aspects and implications of evolution can be satisfactorily explained on this basis alone. Any scientific approach to the problem requires that there be a mechanism which explains why evolution took the pathway it has instead of some other *possible* or *imaginary* pathway. Natural selection undeniably supplies such a mechanism, but is this necessarily the only one which permits such "choice" of pathway, and are not some of the apparent "choices" purely imaginary ones that can be ruled out on other grounds? The question might be put in another way: To what extent has the course of evolution been determined by strictly physical factors that have permitted no exercise of natural selection, and to what extent have the former set the limits within which the latter might act? Can the two things be clearly separated, and what errors of interpretation may be made if we are unable to do so? These are among the questions that will be posed—I do not promise to answer them—in the following pages.

To make even a start in this direction, one needs to view evolution in its entirety—the history of the universe. As in human history, itself a part of evolution, each stage in evolutionary history has depended upon the stage that preceded it, and it is possible to err in interpretation if one studies isolated periods without due regard to the influence of earlier ones. Most works dealing with organic evolution focus attention on its strictly biological aspects, disregarding to a great extent those physical factors that have determined the basic pattern along which living systems could subsequently develop. To understand these factors and restrictions, one needs to go back to the origin of life, and beyond into the domains of terrestrial and cosmic evolution. This is shifting ground, dangerous for the tyro, and caution limits the biologist to a general survey from relatively safe vantage points which may be expected to survive any less than major cataclysms. Yet perhaps it is this limited and necessarily superficial kind of reconnaissance which may be most helpful in the interpretation of later evolutionary events. For the evolutionist may in his cursory exploration stress relationships which are of less importance to the specialist whose approach to those aspects immediately important to his own investigations is more reliable and exact. At least, such a rationalization gives me the temerity to undertake an exploration into the remote past; where I will be, more often than not, venturing beyond the limits of my proper discipline.

The suggestion that life processes are only interpretable in terms of their history will not startle the evolutionary biologist, although he may not be prepared to trace this idea as far back as is attempted here. The physicist or chemist may be less ready to admit this generalization, yet his failure to do so may lead him either to minimize the complexity of the problem of explaining life in physical and chemical terms, or to the opposite extreme which merges with mysticism. The need for taking into account this time dimension of living systems is one of the themes which recurs in the following pages. A good many of the properties of living organisms that appear unaccountable in terms of the inorganic world may stem from events which would be highly improbable within our modern frame of reference, but which, when we take into account the lavish amount of time for them to have occurred in, become much less improbable. The same argument might apply to many properties of the non-living world as well. To appreciate the importance of time in the evolutionary process, one needs to grasp as well as he can the tremendous extent of it which stretches between us and the origin of our planet, and to gain an idea of the relative positions of evolutionary events therein. To orient the reader in this regard, Chapter II summarizes briefly the chronology of earth history and the basis of absolute measurement of evolutionary time. It is said there that life probably originated sometime between two and four billion years ago. But how is the exact moment to be determined?

Perhaps the only way would be to build a time machine—say, an improved and expanded version of that described by the late Mr. H. G. Wells—load it with a group of representative scientists: geneticists, physiologists, chemists, paleontologists, physicists, varieties of morphologists and systematists, or extend the list as you will, and in this happy company travel back at a tempered pace over the four billion years of the earth's history. This would surely be the proper way to reach an understanding of the course of evolution and to pick out the exact point at which life appeared on the planet. Or would it? Although the voyage would no doubt be an instructive one, I think the travelers should be prepared for wide divergence of their opinions, even with the panorama unfolding before them. Each would, I feel sure, be looking for life's origin through different glasses, his criteria depending largely upon the background of his approach to the problem. Conceivably, choices of the point of origin might differ by many millions of years. Not having the facilities at hand for the construction of the time machine my only recourse has been to the expressed opinions of colleagues from various fields; these and my own uncertainty have led me to the above prediction. So, although Chapter X will deal with some of the problems involved in the origin of life, no date and no definite sequence of events will be proposed.

Within our short span of life we are continually aware of the irrevocable passage of time—aware that the same events never exactly repeat themselves whether we wish or no. Viewed in perspective, evolution is characterized by the same one-wayness in time, occasional statements as to its reversibility notwithstanding. It would be useful to us, as evolutionists, if there were some measure of this one-wayness of events. Science offers only one widely general principle which seems applicable; the *second law of thermodynamics*. One way of stating this law is to say that all real processes tend to go toward a condition of greater probability. Sir Arthur Eddington showed insight into the bearing of this law upon our problem when he described it as "*time's arrow*." This implies that the second law of thermodynamics points the direction of all real events in time, although giving no indication of the speed with which they happen. It should be tempting, then, to explore the relationship between time's arrow and organic evolution.

Few, if any, physical scientists would hesitate to apply the second law of thermodynamics to the evolution of the nonliving world; yet even here its applicability may be worth examining. For the second law is in a sense an empirical and pragmatic law which owes its acceptance to the fact that it has worked whenever it has been put to test. The second law can be tested by setting up a self-inclusive system, deducing the changes that should occur, and accurately measuring these changes to see if they agree with prediction. In a sense, we may be accused of rigging the data to obtain agreement, but the fact that we have never failed to obtain it encourages our belief that we deal with a universal principle. Before any claim of a failure of the second law of thermodynamics with regard to any aspect of the nonliving world could be taken seriously, there would have to be absolute assurance that the system involved had been properly set up for examination.

There have been numerous successful applications of the second law of thermodynamics to different aspects of living systems; these encourage the belief that this principle also applies there in a more general sense. Nevertheless, there are from time to time assertions that living organisms manage in some way to violate this principle. In such instances it does not appear that the system has been set up in such a way that it would be possible to reach the conclusion implied, but such statements are likely, because of their dramatic character, to have unwarranted influence on general thought. In Chapter VII living systems as a whole are treated as a thermodynamic system, and no basis is found for claims that the second law of thermodynamics is not obeyed. At other points in the book there will be additional attempts to examine the relationship of this principle to organic evolution, and since many readers may wish an introduction to the subject, Chapter III is devoted to its application to chemical reactions; the kind of application that needs to be made to living organisms as well.

In Chapter III, and from time to time thereafter, I shall resort to the use of analogies and models, with the hope of simplifying certain concepts for the uninitiated, or where physical knowledge is inadequate for a complete description of living systems. In formulating such models, I shall try to keep in mind a statement I once heard made by the physiologist Sir Archibald Vivian Hill, to the effect that one should not construct a model too near to reality, lest he mistake the model for the thing itself. However well I may follow this astute advice, I trust the reader himself will have no difficulty in distinguishing between the model and the real. I shall attempt to choose the models so that the nonspecialist will not be grossly misled, but may be able to gain a reasonable impression of the "forest" without being confused by the "trees." The sophisticate will, of course, recognize the weaknesses, and know how to overlook them while giving his attention to the general argument.[1]

Many things treated in this book have implications within that fascinating terrain where biology and philosophy meet. Indeed, such things alone can make such a treatment worth the undertaking. It seems better, though, to bring them together in the final chapter after the argument has been presented, rather than to anticipate at this point.

[1] But am I too sanguine here, and may not my analogies and models rise up to haunt me ever after?

II · THE CHRONOLOGY OF EVOLUTION

"*Ce qui est admirable, ce n'est pas que le champ des étoiles soit si vaste, c'est que l'homme l'ait mesuré.*"—ANATOLE FRANCE

THE dating of evolutionary events is based on the reasonable postulate that the existing sedimentary rocks of the earth's surface were formed in layers, one after the other. They were formed in different ways, however, and their relationships have been disturbed by various changes, some catastrophic, some gradual. Hence, their relative positions often fail directly to reveal their chronological order. The arrangement of these layers according to a logical time sequence has called for fine detective skill as well as laborious investigation on the part of geologists and paleontologists examining the composition, structure, and fossil content of the rocks as well as their relative positions. This arrangement of the rocks according to their order in time —which provides an outline of evolutionary events—was possible without any definite measure of the lengths of time involved in forming the various layers. Indeed, the basic features of this outline were established long before an accurate method of timing was available. Early estimates of the relative time elapsed in the building of given formations, as compared to others, have proven quite accurate in many cases; but early estimates of absolute time were ridiculously short compared to the values that are now accepted.

Whereas early estimates placed the age of the earth at fifty to one hundred million years or even less, the newer chronology, based upon the "radioactive" method, shows some of the rocks of the earth's crust to be at least as old as four billion years.[1] The age of the earth is now thought to be nearly five billion years. The recognition of this vast

[1] The term *billion* used in this book is the "American" billion, 1,000,000,000 $= 10^9$.

stretch of time available for evolution makes a great difference in the perspective with which this process may be viewed. Much more ample time is allowed for the achievement of given steps than could have been imagined in Darwin's time, or even until very recently.

The radioactive method for determining the age of rocks takes advantage of certain nuclear reactions; for example, the formation of lead from uranium by radioactive disintegration,

$$_{92}U^{238} \rightarrow {}_{82}Pb^{206} + 8_2He^4 + 6_{-1}e^0 \qquad \text{(II-1)}$$

This is a nuclear reaction and is not to be confused with chemical reactions of the usual type, which are treated in the next chapter. It may be read: one atom of uranium 238 goes to one atom of lead 206, eight atoms of helium, and six electrons. The symbols, U, Pb, and He, indicate the chemical species, and the subscripts to the left are the atomic numbers. All atoms having the same atomic number, which represents the number of charges on the nucleus of the atom, have the same chemical properties, so their presence in the above equation is redundant since the symbol for the element says the same thing. There will be more to say about atomic numbers in Chapter VI, but for the present the superscripts to the right of the chemical symbols focus our attention. These are the *mass numbers* (for most practical purposes equal to the atomic weights), which indicate in the above equation that only one isotope of the particular chemical species is concerned. Isotopes are elements having the same atomic number and, hence, the same chemical properties, but different mass numbers and molecular weights. The symbol $_{-1}e^0$ represents the electron, which has a single negative charge and zero mass number.

Lead may also be formed from uranium by the reaction

$$U^{235} \rightarrow Pb^{207} + 7He^4 + 4e^0 \qquad \text{(II-2)}$$

but different isotopes are involved. Lead is also formed from thorium

$$Th^{232} \rightarrow Pb^{208} + 6He^4 + 4e^0 \qquad \text{(II-3)}$$

The isotopes Pb^{206}, Pb^{207}, and Pb^{208} constitute what is known as "radiogenic" lead. Another isotope Pb^{204} present in small quantities in some rocks is called "non-radiogenic" lead because it is not formed by any radioactive process.

The above reactions (II-1, 2, 3) take place at very low rates, which have been accurately determined by the rate of ejection of electrons. For example, the time required for half of a given quantity of pure uranium 238 to change to lead 206 is 4,560 million years. Knowing these rates of radioactive decay, the minimum age of a given rock may be determinable from the quantities of the various isotopes present. Some of the isotopic measurements are more reliable than others; the best methods may serve as checks against each other in some cases. The earliest determinations by the radioactive method were based on what amounted to the assumption that only reaction II-1 took place. The introduction of isotopic analysis and the continual perfection of methods have made revision of age estimates necessary from time to time, but there has never been any question that the age of the oldest rocks are to be reckoned in billions of years.

Other radioactive methods have yielded results which are in general agreement with those involving determinations of lead. Helium is formed in the above reactions and its measurement has also been employed, but being a gas, it is likely to escape from the rocks, and this has introduced uncertainty into the application of this method. It has been applied with success, however, to iron meteorites, which have been assigned an age close to that estimated for the earth's crust. Quite recently another nuclear reaction has been successfully employed, the decay of the element rubidium to strontium.

$$_{37}Rb^{87} \rightarrow {}_{38}Sr^{87} + {}_{-1}e^{0} \qquad \text{(II-4)}$$

These different methods yield somewhat different estimates of the ages of various rocks, but on the whole they are in good agreement.

Within recent years the dates of the earlier rocks have been generally pushed backwards, as the result of new methods and more critical application of older ones. It seems not unlikely that some of the present values may be increased still further. These changes have not decreased confidence in radioactive dating, but have called for more tentative acceptance of dates until further study brings general agreement.

The newer values for the age of the earliest rocks indicate the earth to have been in existence well over the three billion years that was generally accepted a few years ago for the age of the universe. That figure, which must now be regarded as a rough approximation, was based on the apparent rate of expansion of the universe. It is generally agreed on the basis of such estimates that the universe is not over a few billion years old. The concept of the expanding universe has itself been challenged, however, and it seems we shall have to wait for explanation of these and other discrepancies until cosmological theories become stabilized. The problems dealt with in this book are little affected by events outside our solar system and its evolution, so that the above and other cosmological uncertainties need not be of direct concern.

The story of the earth during the past few billion years is to be sought principally in its surface rocks. At some early date the earth's surface was divided into areas of dry land and areas of ocean. It is probable that the areas and outlines of the continental masses have been for a very long time much the same as at present, although the exact size and shape of the dry land area has changed with periods of elevation and submergence of portions of these masses. During periods of submergence, layers of sedimentary rocks were formed, only to be folded and broken during later periods of adjustment of the earth's crust. Intrusion of molten material from deeper layers of the earth into and onto the surface has brought about the metamorphosis of sedimentary rocks as well as the formation of other types. Erosion of the surface by water, and lesser factors such as wind and glaciation, have sculptured the uplifted mountain masses, creating the landscapes characteristic of various geological periods. These processes will be discussed in a little more detail in Chapter V.

For the moment it is enough to say that the traces left by surface changes recurring in a more or less cyclical fashion permit the recognition and dating of events, which in sum constitute the history of the earth's crust before the appearance of a reasonably continuous fossil record. From the beginning of the fossil record we have a powerful tool for unraveling geological history, as well as a record of the evolution of living things themselves. This continuous record begins about half a billion years ago, although there are "sporadic" occurrences of fossils in rocks that go back a few billion years farther. Presumably the first living organisms possessed no hard parts, were too small, or were devoid of other characteristics which might have permitted them to leave their imprints in the rocks, so it must be assumed that the origin of life is hidden somewhere beyond the beginning of the fossil record. Life's exact moment of appearance is a matter of the vaguest speculation, and it may be best to think of this event as having occurred sometime about three billion years ago, but probably spread over a goodly period without being sharply definable.

There are so many uncertainties regarding geological events and the fossil record that it would probably be impossible to outline them to everyone's satisfaction, even in the most general terms. But some such outline is needed if we are to approach the evolutionary problem at all, and Figure 1 is introduced for the purpose. This diagram represents the Geological Time Scale, and the order of certain major occurrences in the fossil record and evolutionary history. The dates may be taken as reasonable approximations, in agreement with currently accepted ideas of geological chronology. They may be subject to revision, and it is to be expected that revision will generally be in the direction of greater age, since dating by the radioactive method gives minimum rather than maximum values. Various systems of nomen-

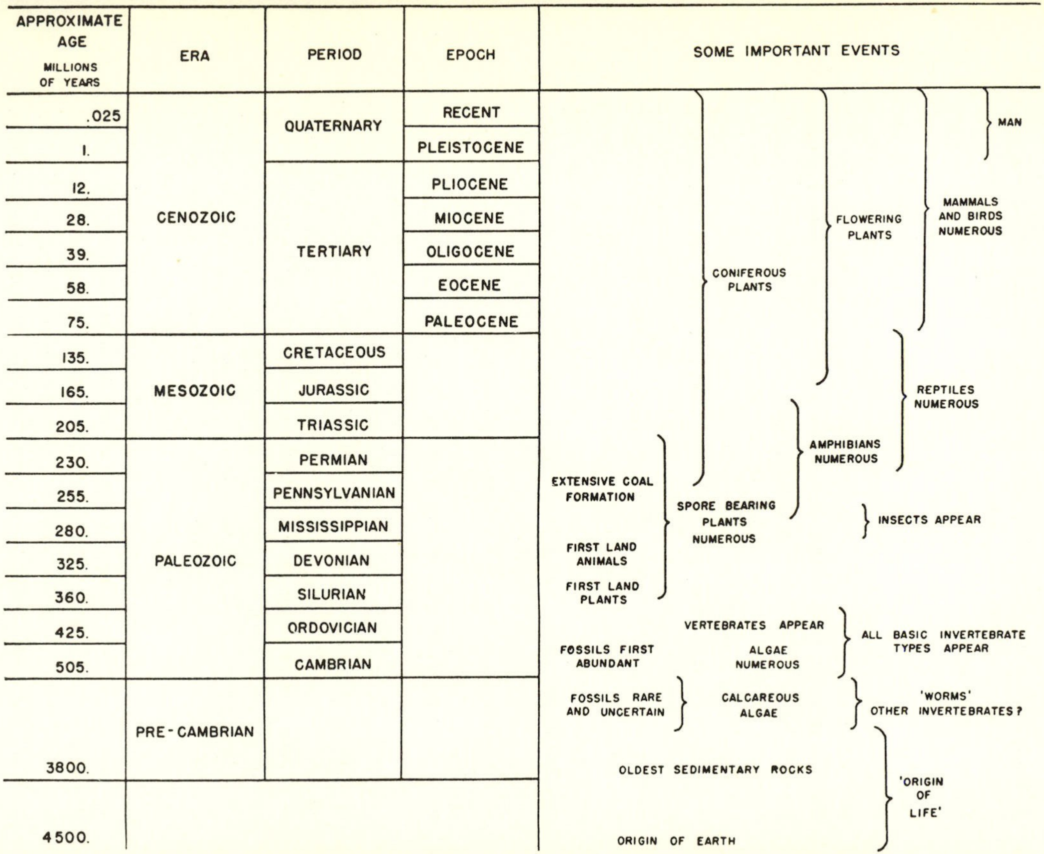

APPROXIMATE AGE MILLIONS OF YEARS	ERA	PERIOD	EPOCH
.025	CENOZOIC	QUATERNARY	RECENT
1.			PLEISTOCENE
12.		TERTIARY	PLIOCENE
28.			MIOCENE
39.			OLIGOCENE
58.			EOCENE
75.			PALEOCENE
135.	MESOZOIC	CRETACEOUS	
165.		JURASSIC	
205.		TRIASSIC	
230.	PALEOZOIC	PERMIAN	
255.		PENNSYLVANIAN	
280.		MISSISSIPPIAN	
325.		DEVONIAN	
360.		SILURIAN	
425.		ORDOVICIAN	
505.		CAMBRIAN	
3800.	PRE-CAMBRIAN		
4500.			

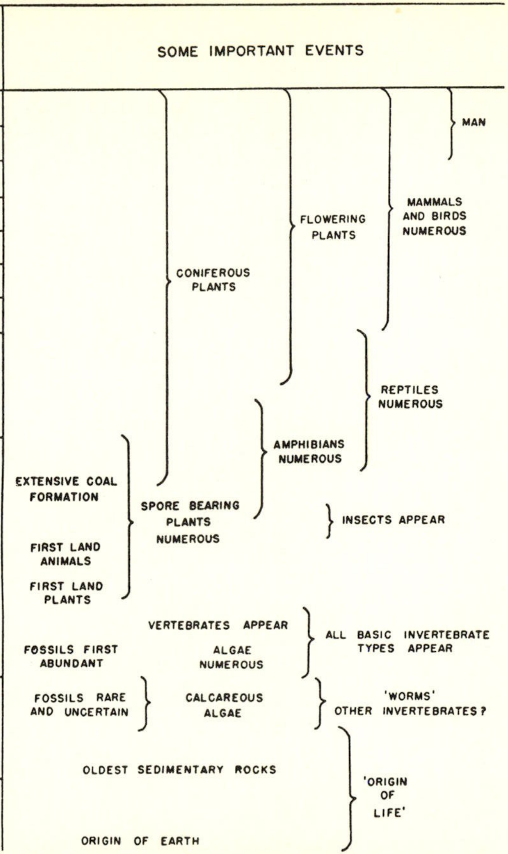

SOME IMPORTANT EVENTS

MAN

MAMMALS AND BIRDS NUMEROUS

FLOWERING PLANTS

CONIFEROUS PLANTS

REPTILES NUMEROUS

AMPHIBIANS NUMEROUS

EXTENSIVE COAL FORMATION

SPORE BEARING PLANTS NUMEROUS

INSECTS APPEAR

FIRST LAND ANIMALS

FIRST LAND PLANTS

VERTEBRATES APPEAR

ALL BASIC INVERTEBRATE TYPES APPEAR

FOSSILS FIRST ABUNDANT

ALGAE NUMEROUS

FOSSILS RARE AND UNCERTAIN

CALCAREOUS ALGAE

'WORMS' OTHER INVERTEBRATES ?

OLDEST SEDIMENTARY ROCKS

'ORIGIN OF LIFE'

ORIGIN OF EARTH

FIGURE 1 The geological time scale, with some important events of the fossil record, and of evolutionary

clature for the geological periods are in use, but the one in the outline is quite generally accepted, so far as it goes. The Pennsylvanian and Mississippian periods correspond, respectively, to the Upper and Lower Carboniferous, as these terms are generally used in Europe. The timing of various events according to the fossil record is not, of course, intended to be either exact or detailed. No attempt is made to assign times of appearance for forms regarding which the fossil record is obscure, for example, the fungi and bryophytes.

The indefiniteness of the record beyond the Cambrian period, more than half a billion years ago, is to be emphasized. Although in the present study we are concerned primarily with the evolution of living things, our interest focusses upon events which took place during the great lapse of time before the beginning of the fossil record; and we must try to follow back to the origin of life through millions of years during which no spoor was left. In this region of shadow one gropes for a few rays of understanding—even the most cautious statement is a speculation. But it is here that the basic patterns for later evolution were determined; where major channels and limits were set. It should be worth the attempt, even though the exploration must be doomed to fail in discovering any more than the most general outlines of events in this remote past.

III · THE ENERGETICS AND KINETICS OF CHEMICAL REACTION

"*. . . of these one of the safest and surest is the broad highway of thermodynamics.*"—GILBERT N. LEWIS AND MERLE RANDALL

CHEMICAL reaction is always associated with thermodynamic changes which determine the direction the reaction takes and how nearly it goes to completion. This is true whether the reaction goes on in a test tube, a geological formation, or in a living system; and must have been true in the infancy of our earth as well as today. Since much of the argument of this book involves thermodynamic relationships, and since many of those interested in evolution may have little familiarity with thermodynamics, particularly as applied to chemical reaction, it may be well to review briefly some of the essentials. For the reader having no knowledge of the subject, I shall introduce, a little later on, a model which may give a picture adequate enough for him to appreciate most of the subsequent argument, or at least serve as an introduction to the somewhat more explicit discussion that follows.

Laws of Thermodynamics

Energy appears in various forms: heat, light, kinetic energy, mechanical work, chemical energy, and so forth. Energy can change its form, but not its quantity—this is a statement of the *first law of thermodynamics*, which until quite recently could be accepted without qualification. We know, now, that matter is another form of energy, but that does not alter this fundamental principle which is also called the law of conservation of energy.

The *second law of thermodynamics* cannot be put in such concise form as the first; it is stated in numerous ways, according to the kind of problem under study. In this book those aspects which bear directly upon the argument will be stressed, to the neglect of others. It is one of this law's consequences that all real processes go irreversibly.

Let us consider a universe, in which the total amount of energy remains, supposedly, constant. Any given process in this universe is accompanied by a change in magnitude of a quantity called the *entropy*, which may be expressed as the amount of heat "reversibly" exchanged from one part of the universe to another, divided by the temperature at which the change takes place. We may imagine the process as being completely reversible, and in this idealized case the entropy change is the minimum possible. In any real process the entropy change is always greater than this theoretical minimum, and the difference between the real and the minimum entropy change measures the irreversibility of the process. All real processes go with an increase in entropy. The entropy also measures the randomness or lack of orderliness of the system, the greater the randomness the greater the entropy; but while the idea of a continual tendency toward greater randomness provides the most fundamental way of viewing the second law, it may not be so apparent in some of the following arguments as are other aspects.

What has been said about the universe also applies to any isolated part thereof, so long as that isolated part—we call it a *system*—does not exchange energy with the rest of the universe. But within any system in the universe which is not so isolated, but which is permitted to exchange energy with its surroundings, the entropy may either increase or decrease in the course of a real process. Living organisms constitute such systems, for they are always undergoing exchange of energy with their immediate surroundings and other parts of the universe. The fact that the entropy can sometimes decrease within such a system is no controversion of the second law of thermodynamics. For if we should expand our system to include all the energy exchange, it would be found that in the larger system the entropy had increased. For instance, to measure the entropy change taking place in living organisms as a whole, it would be necessary to include in our system the sun and some additional portion of the universe, as well as the earth itself. Statements occasionally encountered, to the effect that living organisms do not obey the second law of thermodynamics, are usually based upon the idea that these systems are able to bring about a reduction of the entropy within themselves or their immediate surroundings. Such arguments are beside the point unless the whole thermodynamic balance sheet is taken into account, as their authors seem never to do.

There is another measure which may be applied to the kind of systems in which we are most interested—living organisms—which are comparable thermodynamically to chemical reactions carried out at constant temperature and pressure under thermostated conditions where heat may flow freely to and from the environment. This measure is the *free energy*,[1] which will be discussed at some length in this chapter. The free energy, which represents the maximum amount of mechanical work that can be got out, always tends to decrease in any real spontaneous process, provided there is no interference from the outside that entails the addition of energy other than heat. In most of the applications of thermodynamics discussed in this book, the free energy will focus our attention rather than the entropy.

Since the sum total of all processes that go on in the universe entails an increase in entropy,[2] and since such processes go only one way in time, it may be expected that there is a relationship between time and the second law of thermodynamics. As Sir Arthur Eddington succinctly expressed it, the second law of thermodynamics is "time's arrow." Implicit in this terse phrase is the idea that this law points the direction of all real events in time, and an important corollary is that it does not indicate when or how fast any event will occur; it is time's arrow, not time's measuring stick.

To summarize, then, those implications of the second law of thermodynamics that will be most important in the following pages:

1. All real processes are irreversible, and when all the changes are taken into account they go with an increase in entropy, that is, toward greater randomness or less order.

2. The entropy of a circumscribed system, into or out of which heat energy may flow, may either increase or decrease; but if the temperature of the system is held constant the free energy will always decrease, unless energy other than heat is added from some outside source.

3. The second law of thermodynamics points the direction of events in time, but does not tell when or how fast they will go.

Chemical Bonds

The matter of which both living and nonliving systems are composed is made up, under ordinary conditions, of atoms of around one hundred different kinds (see Figure 7 in Chapter VI). In our ordinary environment at the surface of the earth, atoms are usually combined with other atoms, like or unlike, to form molecules. The atoms are held together in the molecule by forces, electrical in nature, which are

[1] Reference throughout this book is to the free energy of Gibbs, which applies to processes taking place at constant temperature and pressure. The free energy change in such a process is customarily symbolized by ΔF. The free energy of Helmholz applies to processes going on under other conditions, but this function will not be employed herein.

[2] We may think of a reversible process as one that goes at an infinitely low rate, and with a minimum entropy change. A *real* process goes on irreversibly, at a finite rate, and with a definite increase in entropy.

called chemical bonds. In general, a molecule in forming from its atoms undergoes a loss in total energy, the normal molecule representing the configuration which has the least total energy. Thus the chemical bond which unites these atoms may be measured in units of energy. Chemical bonding will be discussed again in Chapters VI and VIII, but here we are more particularly concerned with the overall energy changes in chemical reactions.

Molecules may unite to form other molecules, as, for example, in the reaction of glucose and oxygen to form carbon dioxide and water, which may be represented in the following fashion.

$$C_6H_{12}O_6 + 6O_2 \rightarrow 6CO_2 + 6H_2O \qquad \text{(III-1)}$$

We read this, 1 glucose plus 6 oxygen goes to form 6 carbon dioxide plus 6 water. The letters and subscripts indicate the kind and number of atoms in the molecule. The numbers preceding the symbols for the molecules indicate the numbers of molecules that react.[3] When this reaction takes place there is a rearrangement of the chemical bonds and a loss of total energy, i.e. the substances on the right side of the equation have less total energy than those on the left. But the total change in energy does not determine directly either the direction or the extent to which a reaction goes nor the rate at which it will take place. These matters—direction and rate—are measurable in terms of energy, however, and are related to the total energy, and this introduces other concepts which must now be discussed.

A Model That May Be Useful to Some

The analogy now proposed as a model to illustrate some aspects of the energetics of chemical reaction need not be taken too seriously, but it may serve to introduce the subject to those unfamiliar therewith.

Equilibrium and free energy. Let us imagine a smooth-walled vessel as illustrated in cross section in Figure 2 at I, with a low partition separating the bottom into two compartments 1 and 2. In compartment 1 let us place a number of Mexican jumping beans. For those who may not have been fortunate enough to have played with these in their childhood, I will explain that the "beans" are seeds of euphorbid plants in which moth larvae are enclosed. Occasionally the larva wiggles about in its edible prison causing the seed to move, and if warmed by the clutch of an infant hand the bean may be said to jump. The jumps may not be of great magnitude, actually, but in my remembrance the beans traversed remarkable distances. Anyway, for the purpose of our analogy we will view our jumping beans through child-

[3] Compare this notation with that for a nuclear reaction, e.g. II-1.

hood's eyes and see them leaping about the bottom of the dish. Occasionally one of them, feeling for the moment the urge of excess ambition or perhaps a stomach-ache, will give a more than usually energetic leap, surmount the barrier formed by the partition, and fall into compartment 2. We will imagine this to be a completely random event,

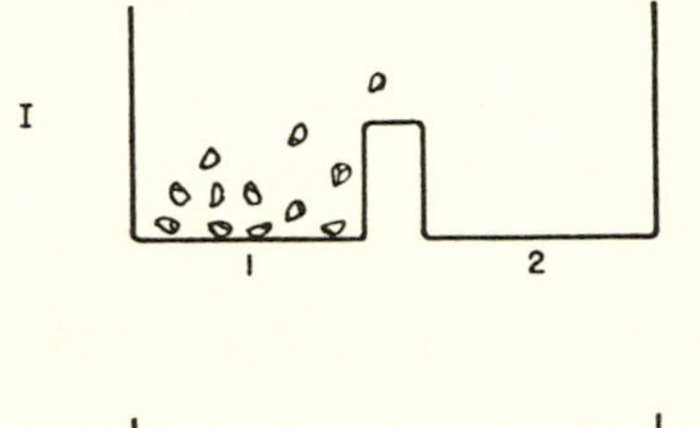

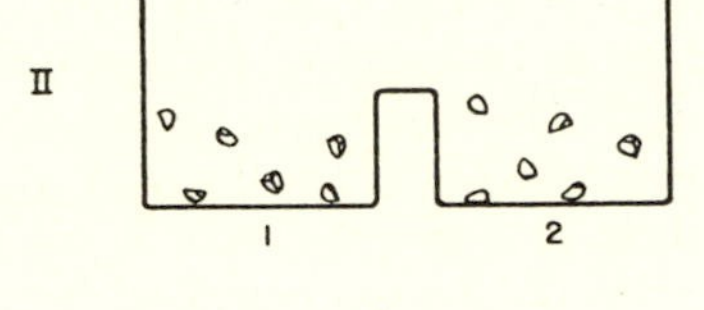

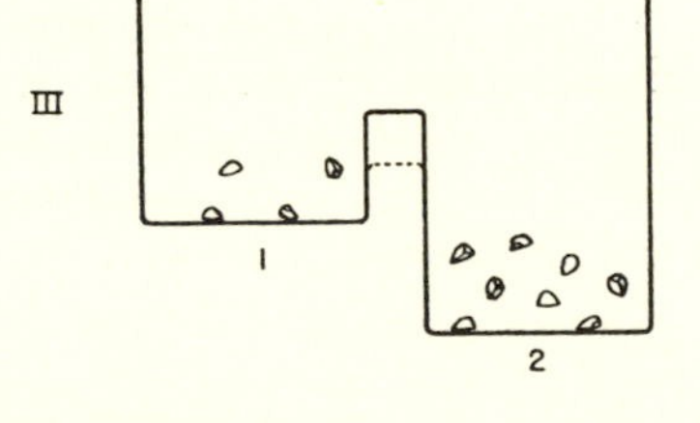

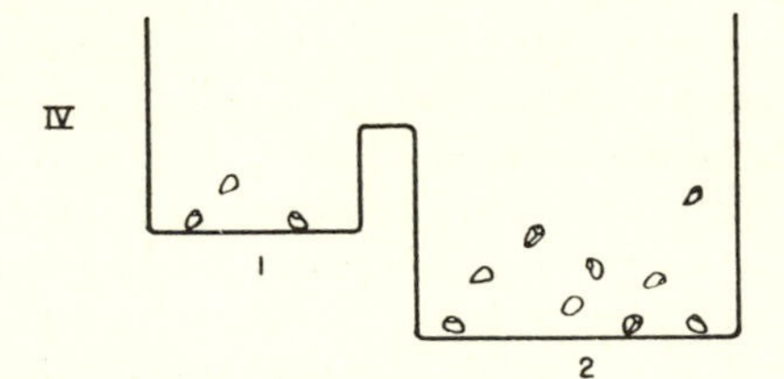

FIGURE 2

happening at irregular, not too frequent intervals. Occasionally one of the beans which has reached compartment 2 will jump back into compartment 1. At first there will be more beans jumping from 1 to 2 because there are more beans in 1, but things will eventually even up so that there are on the average the same number in both compartments, as illustrated in Figure 2 at II, and the same number of jumps per unit time in one direction as in the other. The system will have arrived at *equilibrium.*

But suppose that the level of the bottom of compartment 1 were somewhat higher than that of compartment 2 as indicated in Figure 2 at III. If, now, a number of jumping beans are placed in the higher compartment 1, they will again tend to distribute themselves between 1 and 2; but this time, because a higher jump is required to get from 2 to 1 than from 1 to 2, the latter will occur more frequently than the former. The result is that when equilibrium is reached there will be more of the beans in 2 than in 1, instead of the equal partition found in the previous case.

The jumping bean model may now be compared with a chemical reaction. The beans in compartment 1 represent the reactants, which in the course of the reaction are transformed into products represented by the beans in compartment 2. The levels of the bottoms of 1 and 2 and the top of the partition represent energy configurations which will be discussed shortly. The analogy may seem a little closer if the reactants are thought of as molecules which are transformed into product molecules by some internal rearrangement of their component atoms, without the participation of other kinds of molecules in the reaction. So far as the energy relationships are concerned, however, the analogy will hold just as well for reactions of more than one kind of molecule, for example:

$$A + B \rightarrow C + D$$

where A and B are the reactant molecules which form the product molecules C and D; this is a schematic representation of a reaction like the oxidation of glucose cited above.

If we wait long enough, a condition of equilibrium will be reached in which there is a definite ratio between reactants and products, just as in our model there is an equilibrium distribution of the beans between the two compartments 1 and 2. In the model the distribution of the beans at equilibrium was determined by the difference in levels between the two compartments 1 and 2; and, analogously, in a chemical reaction the equilibrium depends upon differences in energy levels of the reactants and products. The reactant and product molecules may be thought of as jumping back and forth between these two energy levels in somewhat the same fashion as the jumping beans. The ratio of the number of beans in the "product" chamber to that in the "reactant" chamber is comparable to the *equilibrium constant* of a chemical reaction. For example in Figure 2 at II, there are equal numbers of beans in each compartment, so the equilibrium constant is one. In Figure 2 at III there are more beans in the product chamber so the equilibrium constant is greater than one. For a chemical reaction we speak of certain energy levels of the reactants and products as their respective free energies, and the difference between them as the free energy change, or the *free energy of reaction*, which is represented customarily by the symbol ΔF.[4] The free energy change is not exactly represented by the differences in the levels of the compartments in the model, but is a function of the equilibrium constant, which is dependent upon this difference.

Not only does the free energy change determine the ratio of products and reactants at equilibrium, but also the amount of useful work that can be got out of a chemical reaction. Returning to our model, let us suppose that we want to have our jumping beans do some work for us. Referring to Figure 2-III, it is obvious that if a bean jumps from compartment 1 to compartment 2 it loses gravitational potential, and it should be possible to devise some kind of machine for converting a part of this loss to mechanical work, just as work can be obtained with a water wheel from water that loses gravitational potential in running downhill. The maximum amount of work that can be obtained from a chemical reaction is equal to the free energy change ΔF. Of course, all this work could not be got from the reaction any more than can all the energy of a waterfall, but the work obtained could in no case exceed ΔF.

Left to itself, the direction that a reaction takes is toward the condition of equilibrium. In the model, when the levels of 1 and 2 are the same (as II), the balance can be upset at any time by adding beans to one compartment or the other; migration of the beans will then take place until equilibrium is again established. Similarly, a chemical reaction with a small ΔF may be caused to go in one direction or the other by the addition of reactants or products, but it must always go toward equilibrium. On the other hand, if, as in Figure 2-III, compartment 2 is considerably lower than compartment 1, an addition of beans to 1 will result in a considerable shift of beans to 2, but adding beans to 2 will not cause very much migration to 1. In fact, if the difference in level between 1 and 2 is very great there will be practically no chance of any beans getting back from 2 into 1. The oxidation of glucose is a reaction which, once started, will go until there are no glucose molecules left, or at least the probability of finding one may be for all practical purposes disregarded.

It may be said, then, that the *direction*[5] of a chemical reaction is

[4] By convention a negative sign (i.e., $-\Delta F$) indicates that free energy is lost from the reaction when it goes from left to right.

[5] For convenience in thinking we may consider the direction of a reaction as horizontal on the printed page, thus reaction III-1 goes from left to right. The direction of a reaction may also be considered as in the dimension of time.

determined by the sign of the free energy change; the reaction will go spontaneously only in that direction which results in a decrease of free energy, i.e., negative change of ΔF toward zero. The magnitude of ΔF determines the extent to which the reaction will go, and the greater is ΔF the more definitely is the direction of the reaction established.

Appreciable amounts of useful work are to be gained only from reactions with high $-\Delta F$. For example, it is possible to obtain a great deal of work from the combustion of glucose, and for this reason this type of compound is a very important source of energy for living systems.

Rate of reaction. It has been seen that ΔF determines the equilibrium point of a chemical reaction and, hence, the direction in which it will go as well as the amount of useful work it can do. Obviously, ΔF is a very important quantity, yet it tells us nothing about how fast the reaction will go—that is another matter, which may also be explained with the jumping bean model. It is evident that the height of the partition in the model (Fig. 2) has something to do with the rate at which equilibrium is reached. If it were lower, beans making smaller jumps would be able to get over it and in a given time more would do so; if the partition were higher, fewer beans would jump over within a given time, and the time required to reach equilibrium would be increased. The height of the partition finds analogy in the energy of activation, or more properly the *free energy of activation*, of a chemical reaction, which is given the symbol $\Delta F^{\ddagger}$; the corresponding feature has been so labelled in Figure 3, to be more fully discussed later. Changing the height of the partition in the model would not be expected to affect the equilibrium itself, but only the time required to reach equilibrium, and the same is true for a chemical reaction. The magnitude of the free energy of reaction ΔF does *not* determine the magnitude of the free energy of activation $\Delta F^{\ddagger}$ nor vice versa. In some of the subsequent reasoning about living systems (Chap. IX), the importance of this point will appear.

The lack of dependence of the rate of a chemical reaction on the final equilibrium reached is well illustrated by the combustion of glucose. In spite of the great free energy change of this reaction, glucose may remain in contact with the oxygen of the air for long periods without undergoing appreciable combustion, yet under proper conditions, in the human body for example, the reaction goes on quite rapidly. There must be ways, then, of speeding up chemical reactions, and these will now be considered.

Again the jumping bean model is useful. It is the habit of the beans to jump faster and farther the warmer they become, and so if the temperature is raised a few degrees the time required to reach equilibrium should be reduced, because in a given time more beans will jump over the partition from one compartment to the other. The increase in temperature would affect the equilibrium less than it would the rate at which equilibrium is reached. Analogously the rates of chemical reactions generally increase markedly with increase in temperature, whereas temperature has less effect on equilibrium.[6]

There are other ways of increasing the rate of a chemical reaction. It should be obvious from the jumping bean model that, if the energy needed for activation could be decreased in some way, an increase in the rate of reaction would result. This is just what is accomplished in catalysis, which in living systems is brought about by enzymes. The enzymes, which are specific for particular reactions, effectively reduce the energy needed for activation by introducing an alternate path through some intermediate reaction, or a new configuration of the molecule. It is as though, in the model, the partition were cut down at some point to a level represented by the dotted line in Figure 2-III. Most of the transfer of beans would now take place through the gap so made, and the rate of transfer from one compartment to the other would be accelerated. There would be no effect on the equilibrium, however. Correspondingly, enzyme action or any other form of catalysis can alter the rate of a chemical reaction, but cannot alter its direction or its final equilibrium, nor change the amount of work that can be got out of it. Of course, changing the rate does change the power, i.e. the amount of energy per unit time that can be obtained from a chemical process. Thus, if allowed to go on at its ordinary low rate the combustion of glucose would be of no value to a living organism, but when speeded up by enzyme action the energy can be utilized to various ends.

Driving reactions "uphill." Another way of increasing the rate of transfer of beans in the model would be to stimulate individual beans to jump, but without affecting their fellows. Suppose, for example, that with a mirror one could focus a tiny beam of sunlight on an individual bean, and so stimulate it by heating that it would surmount the barrier; then, by focusing the beam on individual beans in compartment 1, one after the other, the rate of transfer to compartment 2 could be increased. Something like this happens in photochemical reactions where the molecules capture light quanta which, being tiny packets of energy, supply energy of activation directly to the molecules. Such activation may result in the speeding of a reaction toward equilib-

[6] For many reactions the shift is very small within the temperature range tolerated by living organisms; but as will be seen in the next chapter the change of equilibrium with temperature may have been an important factor in the evolution of the nonliving world.

rium, or, and this is extremely important, it can actually change the direction of the reaction Referring to diagram III of Figure 2, if individual beans in compartment 2 could be sufficiently heated by the focused beam of sunlight to make them jump over the barrier into 1, more beans might be made to accumulate in 1 than in 2. Analogously a photochemical reaction may be driven "uphill" in an energetic sense. This is made possible because energy is introduced in the form of quanta of energy directly into the molecules. The potential energy of the system is thus increased; and some of this energy can be got back later by letting the reaction proceed along its ordinary course. By such a mechanism (*photosynthesis*) the green plants store the energy of sunlight, which supplies the energy requirements of both plants and animals. This is a matter that cannot be neglected in considering the energetics of living systems as a whole.

A reaction which goes with a decrease of free energy, as all spontaneous reactions do, is called *exergonic.* A reaction in which the free energy increases, as in photosynthesis, is called *endergonic.* Another way of getting energy into an endergonic reaction would be to arrange to have it driven by an exergonic one. The only requirement, from a thermodynamic point of view, is that the exergonic reaction expend more free energy than the endergonic one gains. Thus the "downhill" part is greater than the "uphill," and in sum, the whole process is exergonic. Such *coupled reactions* are of greatest importance in living systems.

The interrelationship between entropy, free energy, and heat of reaction. In I, II, and III of Figure 2, the bottoms of all the compartments are represented as having the same surface dimensions. Suppose that the area of one of the compartments were larger than the other, as illustrated in IV of that figure, where the lower compartment is represented as having a greater area than the higher one. The probability of a bean finding itself in the lower compartment is greater in the case of IV than in the case of III, since the area is larger and there are more places for the beans to occupy. Hence the equilibrium of the beans is shifted by changing the areas of the compartments, so another factor must be taken into account. With analogy to chemical reaction, the ratio of the areas of the compartments would be a measure of the *entropy change*, ΔS, between the reactant and product states. The free energy change, ΔF, would be a function of the equilibrium condition as measured by the ratio of the beans in the two compartments. The actual difference in level between the compartments might be compared to another quantity, the *heat of reaction*, ΔH, which is the quantity of heat exchanged with the environment. All three of these quantities are involved in any chemical reaction, and they are obviously interdependent. Their interrelationships with regard to the jumping bean model will be discussed again in connection with the more exact treatment of the thermodynamics of chemical reaction which follows.

In practice, if we know the free energy change, ΔF, associated with a given reaction going on in a thermostated vessel—or in a living organism under similar conditions—we may predict the direction in which the reaction will go, that is, toward minimum free energy. In making this prediction we may disregard the entropy change, ΔS, which may be either an increase or a decrease; and we may also disregard the heat of reaction, ΔH, i.e. the quantity of heat that flows either into or out of the reaction chamber. If we want to take into account all the thermodynamic changes, however, we must enlarge our system to include this heat exchange with the environment. This is necessary in order to deal with a thermodynamically *isolated system*, that is, one which contains all the thermodynamic changes connected with the particular process. In such a system the direction of change is always toward maximum entropy. We will have to think in terms of free energy, where reactions within living systems are discussed in later chapters. But when living organisms as a whole are considered, reference will be had to the entropy, and it will be seen that a goodly part of the universe must then be included if we are to deal with an isolated system.

Individual molecules. So far the discussion has usually applied to average energy changes of large numbers of molecules, the jumping bean model having been designed to illustrate such a situation. The experimental measurement of thermodynamic properties is made in this way, and classical thermodynamics is based on such reasoning. The newer chemistry based on quantum mechanics treats with the energy changes in single molecules, and this might also have been done with the model by slight changes in assumptions. In Chapter IX it will be suggested that mutations in living organisms may be treated as though they were chemical reactions involving relatively minor rearrangements in large molecules. In this treatment the energy relationships will have to be taken into account and may generally be thought of in terms of Figure 2.

A More Exact Treatment

Let us now undertake a more exact statement of the principles outlined in the last section, which will permit the introduction of some things not covered by the jumping bean model. The points of analogy and of variance with the model will not be pointed out in all cases, but will be left to the curious reader to find for himself.

Equilibrium. Let us consider the reaction of the two hypothetical substances A and B to form two other hypothetical substances C and D,

$$A + B \rightleftharpoons C + D$$

Let us specify that the reaction goes spontaneously at constant temperature and pressure and that exchange of energy as heat may take place freely with the surroundings. Living processes go on under approximately these conditions, and the present treatment may be applied quite directly to them.

Two arrows of different length were used in writing the above expression in order to indicate that the reaction tends to go more toward the right than toward the left. To explain this tendency it is assumed that two opposing reactions which have different speeds occur simultaneously; a forward reaction in which A and B combine to form C and D, and a reverse reaction in which C and D combine to form A and B. The speeds of the two reactions depend on the concentrations of the reacting compounds as well as on specific characteristics of the reactants and products. This situation may be described as,

$$\text{speed of the forward reaction} = k_1[A][B]$$

and

$$\text{speed of the reverse reaction} = k_2[C][D].$$

The symbols for the reactants are inclosed in brackets to indicate concentrations of substances (expressed in moles per unit volume).[7] Specific rate constants characteristic of each reaction are represented by k_1 and k_2.

At the condition of equilibrium, the forward and reverse speeds must be equal,

$$k_1[A][B] = k_2[C][D] \qquad \text{(III-2)}$$

or

$$\frac{[C][D]}{[A][B]} = \frac{k_1}{k_2} = K \qquad \text{(III-3)}$$

The *equilibrium constant* K, which obviously defines the concentration of the reactants at equilibrium, can be determined experimentally in one way or another. The principle involved is generally known as the law of mass action.

It is seen from Eq. III-3 that if K is unity, no reaction will occur when molecularly equivalent quantities of all the substances (A, B, C, and D) are mixed together. If K is greater than unity, the reaction

[7] Strictly, activities rather than concentrations should be used in order to be consistent with the present treatment, but for a general discussion we may think in terms of concentrations. The activity is the product of the concentration and an activity coefficient determinable for the particular situation.

will go to the right under such conditions; if K is less than unity, to the left.

Free energy. The equilibrium constant not only tells what the concentration of the reactants will be at equilibrium, but also shows that if the proportions of the reactants are present in any other concentrations, the reaction will tend to go toward equilibrium. The equilibrium is also related to the amount of free energy change that occurs in going from reactants to products, the relationship being,

$$\Delta F = -RT \ln K \qquad \text{(III-4)}$$

where R is the gas constant, T the absolute temperature, and ln indicates the natural logarithm.[8] The quantity ΔF, the *free energy change*, may be expressed in any desired energy units. The value of ΔF derived from this equation gives the amount of free energy change that must take place in going from a standard condition of the reactants (A, B) to standard condition of the products (C, D). ΔF is usually expressed in calories per mole of one of the reactants. By convention, when the reaction goes spontaneously to the right, ΔF is given a negative value, as seen in Eq. III-4, to correspond to a value of K greater than unity.

Thus, if a reaction has a large negative ΔF, it may be expected to go spontaneously to near completion. If it has a large positive ΔF, it cannot be expected to go spontaneously, but only if energy is added to the system in an appropriate way. If ΔF is small, whether positive or negative, the reaction may be caused to go in one direction or the other by changing the proportions of the reactants to products, but when ΔF is large such change will have little effect on the direction of the reaction. Thus ΔF determines both the *direction* and *extent* of a reaction.

The three following reactions may be used as illustrations:

$$C_6H_{12}O_6(s) + 6O_2(g) = 6CO_2(g) + 6H_2O(l);$$
$$\Delta F^\circ = -688.16 \text{ kg-cal/mol};$$
$$\Delta H^\circ = -673 \text{ kg-cal/mol}. \qquad \text{(III-5)}$$[9]

$$H_2(g) + \tfrac{1}{2}O_2(g) = H_2O(l);$$
$$\Delta F^\circ = -56.56 \text{ kg-cal/mol};$$
$$\Delta H^\circ = -68.39 \text{ kg-cal/mol}. \qquad \text{(III-6)}$$

[8] Absolute temperature is measured in degrees Kelvin. 0°K = −273.1°C. The gas constant $R = 0.00199$ kilogram calories per mole per degree.

[9] The thermodynamic values ΔF° and ΔH° are for temperature 25°C (298.1°K) and pressure 1 atmosphere; they are measured in kilogram-calories per mole. The letters g, l, and s represent the state of the substance; gas, liquid and solid, respectively. In most of the later discussion in this book we will be concerned with the concept of the free energy change rather than with exact values, and the term and symbol ΔF, will be used in a general sense without rigidly specifying the conditions.

$$\tfrac{1}{2}I_2(s) + \tfrac{1}{2}H_2(g) = HI(g);$$
$$\Delta F^\circ = +0.315 \text{ kg-cal/mol};$$
$$\Delta H^\circ = +5.92 \text{ kg-cal/mol.} \qquad \text{(III-7)}$$

The first reaction (III-5) has a very high negative ΔF; hence it will go so far to the right that virtually no glucose and oxygen will be left, but only carbon dioxide and water. The same is true of the second case (III-6). In the third case (III-7) ΔF is positive and so small that the reaction can be caused to go in one direction or the other by moderate changes in the concentrations of the reactants.

Heat of reaction. In the three equations just written above, the new symbol ΔH stands for the heat of reaction. It is measured in the same energy units as ΔF. This symbol represents a quantity of heat energy which the system exchanges with the environment in the course of the reaction; by convention we write $-\Delta H$ if the energy is lost to the environment, and $+\Delta H$ if the energy is gained from the environment. In the former case the reaction is said to be *exothermic* (reactions III-5 and III-6), in the latter case *endothermic* (reaction III-7). The theoretical maximum work that can be obtained from the reaction is not measured by ΔH, but by ΔF, and then only if the reaction goes on reversibly—that is, at an infinitely low rate—is the whole of ΔF obtainable in the form of work. At any finite rate, however, the whole of ΔF cannot be recovered as work. Whatever the amount of work done, the equivalent amount of energy is subtracted algebraically from ΔH, so that if the reaction is exothermic the actual amount of heat given off to the environment is less than ΔH. If ΔF is greater than ΔH (this includes the case of an endothermic reaction) energy is absorbed from the environment to compensate for any work done.

Entropy. Free energy and heat of reaction are related as indicated by the following equation which introduces another term, the *entropy*, symbolized by S:

$$T\,\Delta S = \Delta H - \Delta F \qquad \text{(III-8)}$$[10]

where ΔS represents the change in entropy in the course of the reaction, and T is the absolute temperature. It is obvious from Eq. III-8 that

[10] At the beginning of the discussion it was specified that the reactions take place at constant temperature and pressure, and Eq. III-4 holds strictly for only these conditions. If a reaction goes on with a change in volume, a certain amount of work is done on the environment, equivalent to the product of the pressure P and the change in volume ΔV. In this case the change in total energy ΔE is related to ΔH as follows:

$$\Delta E = \Delta H + P\,\Delta V. \qquad \text{(III-9)}$$

In reactions in living systems it may be assumed that change in volume is negligible and that $\Delta H \cong \Delta E$. The energy of chemical bonds is actually measured by ΔE at 0°K, not by ΔH.

$T\,\Delta S$ is a quantity of energy, since both the quantities on the right-hand side have dimensions of energy. Hence, entropy itself is measured in terms of energy units per degree, i.e., it has the same dimensions as heat capacity.

The entropy itself, S, is a measure of the randomness of the system; the greater the randomness the greater the entropy. One may think of the observable macroscopic properties of any system as the manifestation of the sum of a great many microscopic properties, such, as the energy levels of electrons in atoms, for example, or the vibration of atoms within a molecule, or the momentum of molecules of a gas. A system which has great randomness—high entropy— is one in which these microscopic properties are arranged in a great many different ways. In a system of low entropy, on the other hand, there are only a few possible arrangements of these microscopic properties. We may write for any molecule undergoing chemical change

$$\Delta S = \mathbf{k} \ln \frac{W_2}{W_1} \qquad \text{(III-10)}$$

where W_1 and W_2 are the numbers of possible arrangements at the beginning and at the end of a process involving the entropy change ΔS, and $\mathbf{k}$ is Boltzman's constant.[11] When there is only one possible arrangement

$$S = \mathbf{k} \ln 1 = 0 \qquad \text{(III-11)}$$

Crystals at absolute zero may enjoy only one possible microscopic arrangement, and have zero entropy.

The number of arrangements is proportional to the probability, so we may write

$$\Delta S = \mathbf{k} \ln \frac{P_2}{P_1} \qquad \text{(III-12)}$$

where P_1 and P_2 are the probabilities at the beginning and the end of a given process. Or for a mole of substance, including all the probabilities, P_1' and P_2',

$$\Delta S = R \ln \frac{P_2'}{P_1'} \qquad \text{(III-13)}$$

The interdependence of these thermodynamic properties may be clearer if we substitute from (III-4) and (III-13) in III-8)

$$RT \ln \frac{P_2'}{P_1'} = \Delta H + RT \ln K \qquad \text{(III-14)}$$

[11] Boltzman's constant $\mathbf{k}$ corresponds to the gas constant R. The former is the value for one molecule, the latter for a mole of molecules. $\mathbf{k} = 1.38047 \times 10^{-16}$ ergs per degree.

Or in another form

$$K = \frac{P_2'}{P_1'} e^{-\frac{\Delta H}{RT}} \qquad \text{(III-15)}$$

which shows more directly how the equilibrium constant is determined both by the probabilities of the beginning and end states of the reaction itself, and by the heat exchange with the environment.

We find an analogy in the jumping bean model. The quantities P_1' and P_2' may be thought of as proportional to the areas of the two compartments; ΔH as the difference in level between these compartments; and K as the ratio of the number of beans in the two compartments when equilibrium has been reached.

High temperature is associated with high randomness, and when there is a flow of heat from a warmer to a cooler body the randomness of the two systems, considered together, increases—there is an increase in entropy. The flow of heat may be used to have work done by the system, but work would have to be done on the system to return it to its original condition. In isothermal chemical reactions, on the other hand, heat may flow either to or from the surroundings depending upon the particular reaction. Work may be obtainable from the reaction in either case, for the entropy of the system as a whole increases, due to increased randomness in microscopic arrangement.

A general postulate of the second law of thermodynamics is that in any spontaneous process the entropy must increase toward a maximum, and hence entropy change, ΔS, like free energy change, ΔF, is a measure of the direction of a reaction. It may seem confusing at first to note that in the reactions III-5, III-6 and III-7, the entropy changes are not proportional to the free energy changes; the numerical values of $T\Delta S$ are, respectively, for the three, $+15.16$, -11.83 and $+5.605$ kilo-calories per mole, while the corresponding values of ΔF are -688.16, -56.56, and $+0.315$ kilo-calories per mole. How, then, can both ΔS and ΔF be measures of the direction of the reaction? The answer is that the free energy is such a measure for the reaction itself without respect to the heat exchange with the thermostat bath or the universe in general; quantitatively, ΔF tends to go toward zero. The entropy change, on the other hand, refers to the reaction and its surroundings in an overall sense, and in this sense the direction of the reaction is determined by the entropy change which always goes toward a maximum.

The free energy and entropy are different aspects of the system as a whole. Obviously the conditions and the system under consideration must be understood before attempting to apply the second law quantitatively to any process, including the chemical changes that go on in living systems. In Chapter VII the energetics of such systems will be discussed at some length, and it will be seen that in order to obtain a complete balance sheet it is necessary to journey far beyond the confines of the earth itself.

Reaction kinetics. Another aspect of chemical reaction now presents itself—that having to do with rates of reaction. The free energy change ΔF specifies the equilibrium conditions for a given reaction, which will be reached if the reactants are left to themselves for sufficient time. But this quantity gives no information as to how rapidly that equilibrium will be reached. Glucose may remain in contact with air for a very long time without appreciable oxidation taking place, yet

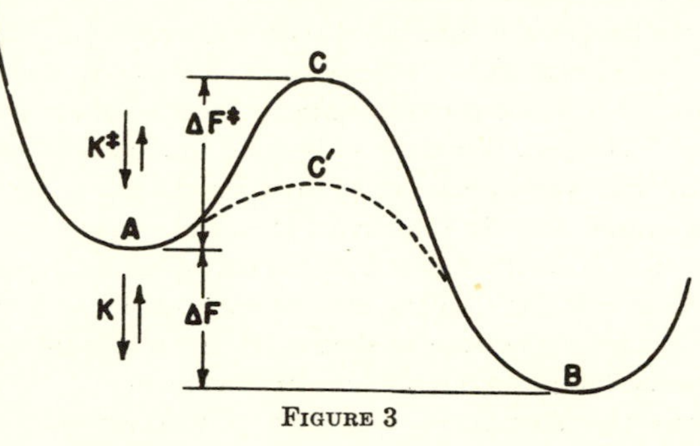

FIGURE 3

reaction III-5 is highly irreversible, having a very large negative ΔF. At ordinary temperatures catalysis of some kind is necessary, and may cause the reaction to go at a relatively high rate. Likewise, in spite of the high ΔF of their reaction, gaseous oxygen and hydrogen (III-6) may remain together indefinitely until a catalyst or a local increase in temperature is provided, after which the reaction may go with explosive violence. Obviously, the free energy of a reaction (ΔF) gives no information about its rate.

Modern kinetic theory pictures the reactant molecule as only able to react if it can achieve a certain energy value, when it is said to be in an "activated state." This situation may be represented as in Figure 3, where the reactant molecules A are represented as in a trough from which they can only reach the product state B, represented as in a lower trough, by passing through an activated state C having higher energy than A. The free energy of reaction may be regarded as a function of the ratio of forward and reverse rates of the reaction, i.e. the equilibrium constant K (Eqs. III-2 to III-4). The reaction may be represented by

$$A \underset{}{\overset{K}{\rightleftarrows}} B \qquad \text{(III-16)}$$

According to the theory formulated by Henry Eyring, a similar equilibrium may be assumed to exist between the reactant molecule and its activated state, which may be represented by

$$A \underset{\longleftarrow}{\overset{K^{\ddagger}}{\longrightarrow}} C \qquad \text{(III-17)}$$

$K^{\ddagger}$ being the ratio of the frequencies with which the molecule is to be found in the two states A and C respectively. The *free energy of activation*, $\Delta F^{\ddagger}$, may then be represented in the same way as ΔF in Eq. III-4.

$$\Delta F^{\ddagger} = -RT \ln K^{\ddagger} \qquad \text{(III-18)}$$

The free energy of activation, $\Delta F^{\ddagger}$, and the free energy of reaction, ΔF, may be represented as in Figure 3 if free energy is measured along the ordinate. The specific reaction rate k, i.e. the number of moles of substance reacting per unit time, may be described by,

$$k = \kappa \frac{\mathbf{k}T}{h} e^{-\Delta F^{\ddagger\prime}/RT} \qquad \text{(III-19)}$$[12]

where $\mathbf{k}$ is the Boltzman constant, h is Planck's constant,[13] κ is a "transmission" coefficient which for most purposes may be regarded as unity, and e is the base of natural logarithms. This equation tells us that the rate at which the molecules react is not determined alone by the magnitude of the energy barrier that must be surmounted, but also by the frequency with which this energy is achieved.

Rates of chemical reactions can be altered in different ways. The rate of reaction is dependent to a considerable extent upon the temperature of the system. As a general rule, rate increases with temperature; and in a good many cases within the range of biological temperatures approximately doubles with a 10°C rise. This rule of thumb is by no means exact, however, there being numerous exceptions. Increase of rate with temperature is interpreted as due to increase in the number of molecules that achieve the required energy of activation within unit time; in fact the determination of the energy of activation is based on the variation of rate with temperature.

Increase in rate may also be accomplished by introduction of a catalyst—the enzymes are biological catalysts. The effect of a catalyst is interpreted energetically as due to the lowering of the free energy of activation, as suggested in Figure 3 by the dotted line, the activated state for the catalyzed reaction being indicated by C′. Since in unit time more molecules may reach C′ than would reach C, the rate of the reaction should be increased. The lowering of the energy of activation

[12] $F^{\ddagger\prime}$ differs very slightly from $F^{\ddagger}$.

[13] Planck's constant, $h = 6.6236 \times 10^{-27}$ erg seconds.

may be brought about by interpolating intermediate steps between A and B, or perhaps in other ways. It is to be emphasized that although the rate of a chemical reaction may be increased by the introduction of a catalyst, this does not alter the direction[14] nor the extent of the reaction. As indicated diagrammatically in Figure 3, the free energy of reaction ΔF is not changed by the catalysis, but only the free energy of activation $\Delta F^{\ddagger}$.

Photochemical reactions. All the above reasoning regarding the kinetics of chemical reaction is based on the behavior of the molecules in a group. Among these molecules the energy of the group as a whole is distributed at any instant in a certain fashion, and the rate of the reaction depends upon the proportion which will achieve some given energy value per unit time. This distribution of energies may be upset when it is possible to add energy from the outside directly to the molecule, as for example when a molecule captures a quantum of radiant energy as in a photochemical reaction. The primary act in such a reaction is represented by

$$M + h\nu = M^*. \qquad \text{(III-20)}$$

where M is the light absorbing molecule which captures a quantum of radiant energy $h\nu$ and so becomes an activated molecule M^*. An activated molecule may participate in reaction and in such case the rate of reaction may be directly proportional to the number of quanta captured per unit time. Not only may reactions which take place spontaneously at a finite rate—that is, have a negative ΔF—be accelerated in this way, but in some cases sufficient energy may be supplied to the molecule to drive the reaction "uphill," in the direction opposite to that indicated by ΔF. An example of this type of reaction, which is of the greatest importance biologically, may be represented thermodynamically by

$$\begin{aligned} H_2O + CO_2 &= \{CH_2O\} + O_2; \\ \Delta F &= +115 \text{ kg-cal/mol;} \\ \Delta H &= +112 \text{ kg-cal/mol.} \end{aligned} \qquad \text{(III-21)}$$

Reaction III-21 represents, thermodynamically, the process of photosynthesis by green plants, about which much will be said later. Although this reaction proceeds in the opposite direction to that expected for a spontaneous (thermal) reaction, this in no way controverts the second law of thermodynamics, since in this photochemical reaction energy from outside the system is added in such a way as to upset the statistical relationships that determine the direction in which

[14] That is, if we regard the direction of the reaction as always toward equilibrium (e.g. see Chapter VII).

the reaction would go spontaneously. This is a very different thing from catalysis, which does not alter the direction of a reaction.

Exergonic and endergonic reactions. The terms *exergonic*, describing reactions that go with decrease in free energy, and *endergonic*, describing reactions that go with increase in free energy, are convenient and now in common use. All spontaneous reactions are, of course, exergonic. Endergonic reactions only occur when free energy is supplied in some way from outside—for example, in photosynthesis, where light energy is changed into potential chemical energy. Under some circumstances an exergonic reaction may supply the free energy required to drive an endergonic one. In such a *coupled reaction*, the total free energy decreases, so that the overall process is exergonic.

Since the original appearance of this chapter (in which the treatment of thermodynamics may be regarded as "classical"), much attention has been given to thermodynamics of irreversible systems and the treatment of steady states in *open systems*. The latter are chosen so that not only the exchange of heat, but also the exchange of other forms of energy can be "neglected," in the sense that heat exchange is neglected in isothermal reactions as treated above. As should become clear in Chapter VII, living systems may be regarded as open systems if we choose, and for some purposes this may prove a useful tool for studying them. If we are to draw up a thermodynamic balance sheet for evolution, we must include all exchanges with the environment. We do not ordinarily do this, but tend to think of evolution as a thing by itself without regard to the larger sun-earth system of which it is a part. We must be careful not to be misled. A general model, outlined in Chapter XIII, may be of help to us in viewing evolution as a process in which order increases but nevertheless obeys the second law of thermodynamics.

IV · THE ORIGIN AND EARLY EVOLUTION OF THE EARTH

"*Ought not a process that for eternity has taken but one direction, to have been completed long ago?*"—JOSIAH ROYCE IN *1892*

"*Instead of the 30 billion years . . . only 10 or 12 billion years.*"
—H. A. BETHE IN *1942*

HYPOTHESES regarding the manner of the earth's origin have come and gone, each enjoying its vogue for a time, only to be replaced by another which more nearly fitted the available evidence. At present we find ourselves in a period of radical change in concepts. The collision hypothesis, which not many years ago was almost accepted dogma, has succumbed to inherent weakness, and in its demise has carried away with it various notions regarding terrestrial evolution. This hypothesis provided the basis for a good deal of thinking about the origin of life. It postulated that the close approach of another large celestial body to our sun resulted in the extrusion of a small portion of the sun's mass, from which were formed the planets of our solar system.[1] The earth, supposed to have been at first gaseous, like the sun, cooled rapidly becoming first liquid and then solid. On this basis it could be assumed that the earth was once made up of the same kind and quantities of materials as the surface region of the sun, about the constitution of which a good deal is known from spectroscopic analysis. Things known about the sun's surface might be applied then to description of the earth at the moment of its birth, and its subsequent evolution be deduced from this starting point. A great weakness of the collision hypothesis lay in its failure to account for the distribution of angular momentum among the sun and planets. A number of plausible alterna-

[1] The impact of newer ideas was only beginning to be felt when this chapter was written, and it contains statements reflecting the collision hypothesis, which should be modified. It may be sometime, however, before the new concepts are brought together in a form that will be generally acceptable, and only limited revision will be made at this time.

tive hypotheses have been proposed which avoid this difficulty, but none of these is universally accepted. The idea that the sun and planets were built up from a common mass of gas, or by the accretion of particles from a cosmic "dust cloud," are among the alternatives.[2] The collision hypothesis permitted the picture of a series of evolutionary events starting with what is known about the chemical constitution of the sun, and ending with what is known about that of the earth. One had to use his imagination regarding the intervening course of events, but the starting point seemed clear. The new theories allow no real certainty as to the starting point. The nascent mass may not have been entirely gaseous, nor have approached the temperature of the photosphere.

Nevertheless, in spite of the artificiality of such an approach, it may still be useful to consider the events that might have taken place in the formation of the earth from a mass having the constitution of the photosphere of the sun. We need to begin somewhere, and so long as our approach is sufficiently general and we do not place great confidence in the specific mechanisms that are postulated, we may not go too far astray. The same general principles are involved sooner or later in any approach to the problem.[3] At any rate, this is the path that will be followed, here, with the hope that it may lead us to a better understanding of factors that may have determined the fitness of this planet as an abode for life.

Whatever the circumstances of our planet's birth, the sun has been of greatest importance to the evolution of life thereon, and a brief account of its constitution would be appropriate for this reason alone. Here, as in many other places, the rise of nuclear physics and chemistry has greatly altered our point of view. Some years ago H. A. Bethe, after examining various nuclear reactions that might provide the energy of the sun, arrived at the following scheme, which appears to be unique in satisfying all the required conditions and is now generally accepted:

[2] Recent hypotheses all assume some form of cloud, whether gaseous or particulate.

[3] The assumption that the cloud from which the earth was formed had at some time an elementary composition something like that of the photosphere serves as a starting point for most reasoning about the problem. It is unlikely, however, that the temperature of the cloud ever approached that of the photosphere, although it may have risen quite high due to adiabatic condensation and radioactivity. Hence those statements which presuppose very high temperatures should be revised. At present, different writers postulate different sequences of events and this leads to different ideas regarding the origin of various constituents. The same factors are always to be invoked, however, e.g. changes in temperature that shift equilibria, loss of light elements, layering due to gravity, etc.; so the content of this chapter need not be misleading unless taken too literally.

$$
\begin{aligned}
&{}_{6}C^{12} + {}_{1}H^{1} \rightarrow {}_{7}N^{13} + h\nu \\
&{}_{7}N^{13} \rightarrow {}_{6}C^{13} + {}_{1}e^{0} \\
&{}_{6}C^{13} + {}_{1}H^{1} \rightarrow {}_{7}N^{14} + h\nu \\
&{}_{7}N^{14} + {}_{1}H^{1} \rightarrow {}_{8}O^{15} + h\nu \\
&{}_{8}O^{15} \rightarrow {}_{7}N^{15} + {}_{1}e^{0} \\
&{}_{7}N^{15} + {}_{1}H^{1} \rightarrow {}_{6}C^{12} + {}_{2}He^{4} \qquad \text{(IV-1)}
\end{aligned}
$$

The mass number (approximately equal to the atomic weight) of the isotope is indicated in each case by the superscript to the right of the chemical symbol, the subscript to the left is the atomic number, as for the nuclear reactions described in Chapter II.[4] ${}_{1}e^{0}$ represents a positron, a particle having the mass of the electron and unit positive charge, and $h\nu$ is a quantum of gamma radiation of very short wavelength, which will be absorbed almost immediately by electrons and reemitted as smaller quanta. The positrons combine with electrons to give gamma rays. After a long time, during which many absorptions and reemissions take place, a small part of the energy of these gamma rays, transformed into radiation of much longer wavelength, is emitted from the surface as sunlight.

The net result of this reaction is the transformation of hydrogen to helium, accompanied by release of energy. This energy should last for several billion years, and, contrary to earlier belief, the sun may be getting hotter instead of cooler. If this is true, life—barring some intervening catastrophe—may look forward to extinction in several billion years by intense heat, instead of by intense cold as was once anticipated. In the meantime life may continue, as today, to sustain itself by means of the energy of sunlight. Since the nineteenth century there has been a tremendous extension of our perspective of time, both into the past and into the future, beyond anything that then seemed reasonable. The two quotations at the beginning of this chapter suggest a shift in point of view, affecting philosophy as well as science.

The reactions described by the above equations take place in the central part of the sun where the temperature is estimated to be about 20,000,000°K. Peripheral to this, where the temperature is lower, nuclear reactions take place which result in the formation of other elements, and in the surface regions a variety of these are present. Only a very superficial region of the sun can be directly observed, since it becomes opaque quite near the surface due to the density of its gaseous components. But the outer envelope or *photosphere*, the bright disk seen with the naked eye, may be examined with the spectroscope, and a good deal of information obtained about its constitution. It is

[4] A brief discussion of the structure of atoms appears in Chapter VI.

found that the maximum emission is at wavelength 0.48 μ, at which wavelength a perfect radiator ("black body") at about 6,000°K emits its maximum radiation, indicating that the photosphere has about this temperature.[5]

I shall assume tentatively for purposes of discussion that the mass from which the earth was formed had approximately the temperature of the sun's photosphere and was in the gaseous state. Once separated as an isolated body, this mass should have cooled rapidly by convection in the interior and radiation from the surface, first becoming molten and later solidifying at the surface to form a crust. Jeffreys estimated that a mass the size of the earth would have reached a temperature of 1,500°K—at which a solid crust might be expected to begin to form—within about 15,000 years. In about 25,000 years the temperature of the surface would have reached very nearly that of the earth at present. Thus the initial period of rapid cooling would have occupied only an insignificant part of the total age of the earth. These estimates are based on the conditions assumed for the collision theory, however, and so are uncertain. Nevertheless they indicate that whatever the mode of its formation, the earth was not for long at a very high temperature. Events of great importance to the future history of the earth could have taken place, however, during an initial period of rapid cooling.

Spectroscopy also permits the identification of many elements in the sun's photosphere; the number known should increase as methods are developed for studies from outside the earth's atmosphere, which restricts observations in the ultraviolet and infrared regions of the spectrum (see Figure 5). The atoms of each element absorb and emit only radiation of a limited number of wavelengths, the position of which in the spectrum identifies the element. Thus, for example, gaseous sodium absorbs the wavelengths 0.5890 μ and 0.5896 μ, and hence when the spectrum is viewed through a layer of this gas, two dark "lines" appear in the yellow region. Other dark lines reveal the presence of other elements in the sun. Estimates of the relative quantities of the more plentiful are presented in Table 1. That the sun is not a unique kind of body in the heavens is indicated by its

[5] The term "black body" is applied to a theoretically perfect radiator. The wavelength of maximum emission λ_{max} for such a radiator is given by Wien's law:

$$\lambda_{max} \times T = 0.2885 \text{ centimeter degrees}$$

where λ is measured in centimeters and T is the absolute temperature in degrees Kelvin (°K). The accepted value for the sun's temperature is 5,710°K. See Figure 4 for the spectral distribution of sunlight.

having the same surface composition as all the stars of the main sequence. Thus, without restriction as to theory of origin, the photosphere of the sun serves as a reasonable basis for comparison and for considering the earth's evolution, since it represents to a certain extent stellar matter in general.[6]

At the high temperature of the photosphere most of the elements are in the atomic or ionic state, but some molecules are present also, particularly in the areas of sunspots, where the temperature may be much lower than in the rest of the photosphere.[7] These molecules, too, may be identified spectroscopically, by their characteristic "band" spectra. The presence of molecules in the photosphere shows that chemical reactions may occur at the temperature and pressure existing there. Such reactions are limited, however, to a small number of combinations of atoms to form diatomic molecules, as for example,

$$C + N \rightleftarrows CN. \qquad \text{(IV-2)}$$

Equilibrium constants may be very small at such high temperatures. For the above reaction K is about 10^{-12} at the conditions of the photosphere, which means that, proportionate to the quantities of C and N atoms present, only minute quantities of the product would be formed. Nevertheless, enough CN is present in the atmosphere of the sun to be identifiable spectroscopically; it should be remembered that the atoms of C and N tend to combine constantly to form CN, even though there is a much greater tendency for the CN molecules to break up into the constituent atoms, so that at equilibrium there are always some CN molecules present. Numerous other molecules are found in the photosphere, likewise in very small proportions: CH, NH, OH, MgH, SiH, SiF, C_2, O_2, MgO, CaH, BH, SiO, AlO, ZrO, MgF, and SrF, have been clearly identified, and the presence of others has been claimed. Some, such as N_2, may be present but are not detectable because their spectra fall in the ultraviolet not observable through the earth's atmosphere.

Cooling of such a mass as we imagine would result in changes in chemical equilibria in the direction represented by higher values of K, tending to increase the proportions of those molecules present. In addition, other reactions would become possible and other species of molecules appear. Rates of reaction would be even more affected by changes in temperature, tending to favor certain reactions over others.

[6] Outside the photosphere is another very tenuous zone which, as indicated by the state of its atoms, must be at a much higher temperature than the photosphere itself.

[7] Russell, H. N., *Astrophys. J.* (1934) *79*, 317–342; Babcock, H. D., *Astrophys. J.* (1945) *102*, 154–167.

Thus, as temperature continued to fall, it might be supposed that, in a chemical sense, complexity would breed complexity, since the more species of molecules were present, the more could form. The formation of new species of molecules would take place at the expense of some of the existing ones, some of the species present at a given temperature tending to disappear as cooling continued. The general picture would be the same if the nascent earth had had a considerably lower temperature than that of the photosphere, although higher than now.

Thus far, the cooling mass has been pictured as a homogeneous system, but this would hardly have been the case for very long. The lighter atoms would tend to move away from the center of the mass more rapidly than the heavier ones, resulting in a certain degree of layering. A good many of the lightest atoms would escape into space. To escape from the present earth's gravitational pull, a molecule must attain a velocity of 11.3 kilometers per second while moving in a direction normal to the surface, and such velocities would be reached by many of the lighter atoms. The average velocity of the atoms of a gas is inversely proportional to the mass of the individual atoms, and directly proportional to the square root of the absolute temperature. At the temperature of the photosphere the *average* velocities of the two lightest species of atoms, hydrogen and helium, are about 2.4 and 1.2 kilometers per second respectively, but some of the atoms would reach much higher velocities. Under the conditions assumed, these two gases should have been virtually lost from a body the size of the earth within a relatively short time, had the temperature not fallen rapidly.[8] The present composition of the earth shows that if hydrogen and helium were once present in proportions comparable to those in the sun's photosphere, they have been lost in great quantities, as the comparison in Table 1 shows.[9]

The difference in the proportions of helium in the sun and in the earth is, however, very much greater than is the case for hydrogen. It would seem that more helium has been lost proportionately than hydrogen, in spite of the fact that the atoms of the latter reach twice as high velocities as do those of the former. This difference might be explained in part at least by the combination of hydrogen with heavier elements, for example with silicon to form the molecule SiH which

[8] See Jeans, J. H., *Dynamical Theory of Gases*, Cambridge (1925), Cambridge University Press, 4th ed., or for a brief account, Jones, H. S., *Life on other Worlds*, New York (1940) Macmillan.

[9] As will be seen in the next chapter, the composition of the earth presented in the Table 1 is based on assumptions which may be open to question, and the values given should in most cases be regarded only as tentative estimates. With regard to the elements in question at this point, however, the data should be adequate to support the argument.

exists in the sun's photosphere and whose average velocity is only about one fifth that of hydrogen. Temporary combination in such molecules could have helped to "save" some hydrogen from escape, until lower temperatures were reached at which more stable compounds were formed and held. Helium, on the other hand, forms no such combinations with other atoms, and it would not be suprising if it had been entirely lost. Hydrogen is virtually absent from the atmosphere, but is present in combination in the earth.

Table 1. Proportion of Elements in Earth and Sun.

Element Name	Element symbol	Atomic weight	Atomic per cent[1] in sun's photosphere[2]	Atomic per cent in earth[3]
Hydrogen	H	1.00	81.76	2.7
Helium	He	4.00	18.17	(10^{-9})
Carbon	C	12.01	0.003	0.1
Nitrogen	N	14.01	0.01	0.0001
Oxygen	O	16.00	0.03	48.7
Sodium	Na	23.00	0.0003	0.7
Magnesium	Mg	24.32	0.02	8.2
Aluminum	Al	27.97	0.0002	2.4
Silicon	Si	28.06	0.006	14.3
Sulfur	S	32.07	0.003	0.7
Potassium	K	39.10	0.00001	0.1
Calcium	Ca	40.08	0.0003	2.0
Titanium	Ti	47.90	0.000003	0.3
Vanadium	V	51.95	0.000001	—
Chromium	Cr	52.01	0.000006	0.05
Manganese	Mn	54.93	0.00001	0.08
Iron	Fe	55.85	0.0008	17.9
Cobalt	Co	58.94	0.000004	0.9
Nickel	Ni	58.69	0.0002	1.4
Copper	Cu	63.54	0.000002	0.005
Zinc	Zn	65.38	0.00003	—

[1] I.e., per cent of total atoms.
[2] From data for "percentage volume" in solar atmosphere as quoted from D. H. Menzel by Goldberg, U., and Aller, L. H., *Atoms Stars and Nebulae*, 1943, Blakiston.
[3] Based on data presented in Table 2.

Table 1 shows that in general the lighter elements have tended to be lost from the earth, as would be predicted because of the greater relative velocities of their atoms; but there is a wide variation from any general rule based on the atomic weights alone. Perhaps this variation might be accounted for to a considerable extent on the basis of respective "affinities" for other atoms, that is, the possibility of combination with other species to form heavier particles that would have lower velocities than the separate atoms. For example, carbon, nitrogen, and oxygen have low atomic weights, but nitrogen seems to have been largely lost, whereas a much greater proportion of carbon and oxygen have been retained. This could be accounted for by the greater readiness with which the latter two elements combine with other atoms to form molecules.

Because of the differences in their velocities, the atoms and molecules might have tended to form layers, the heavier concentrating toward the center of the mass, the lighter toward the periphery; and as the outer layers cooled, diffusion outward should have been slower than movement inward, thus enhancing the layering. This factor could have had a considerable role in determining the amounts of various molecules that were formed. For example, other things being equal, combinations among the lighter elements concentrated toward the cooler periphery might have been greater than combination of the heavier elements at higher temperatures prevailing toward the center. Under such conditions combination between the very heavy and the very light elements might have been relatively rare.

In such a mass, iron should begin to liquify at its boiling point, 3000°C, this occurring first at the outer periphery of the mass, with molten droplets of iron falling rapidly toward the center because of the increased effect of gravity upon them. In keeping with such a picture, the iron in the earth might be expected to be concentrated in a central core, together with smaller quantities of nickel and cobalt, which have atomic weights and boiling points near that of iron. The important elements Si, Al, Mg, and Ca would also combine as oxides but would probably be, for the most part, still in the gaseous state after the iron had liquified.

With further cooling, a surface crust of solid iron could begin to form; but by this time many of the other elements and compounds should have liquified or even solidified, and precipitated as droplets or solid particles onto the outer surface of the iron core. For example, some of the oxides of silicon have boiling points lower but melting points higher than those of iron, so that these compounds could have formed a solid crust before the iron itself began to solidify; that is, they would have been in the gaseous state at the time the iron liquified, but would have liquified and then solidified while the iron was still liquid. These oxides, having much lower specific gravities than iron, should have tended to remain on the surface; the same should apply to other compounds of the lighter elements. A solid crust over the liquid iron core would probably have broken up from time to time as the core cooled and contracted, portions being engulfed and remelted, thus enhancing the variety of compounds as cooling progressed.

As the result of the operation of all these factors, the earth could have had, after still further cooling, a general make-up resembling that it is thought to have today. There should have been a core at the center consisting principally of elementary iron, surrounded by a mantle composed chiefly of materials of lower specific gravity. This mantle should have varied in density, decreasing from the inside outward. Because of their atomic weights, the quantities originally in the sun, their affinities for oxygen, and the boiling and melting points of their oxides, the elements Na, Mg, Al, K, Si, and Ca should have been present in large quantities in the mantle, together with oxygen and iron, the former combined with all these elements. Outside this mantle there should have been an atmosphere of gases. In a general way the constitution of the earth is similar to that described. Probably one would not arrive at the above description, however, without these ideas of the present structure of the earth, and hence, in a way, it must be regarded as only a rationalization aimed at harmonizing supposed past events with known present structure. But perhaps the important thing is that it is possible to make such a rationalization in the absence of so much vital information.

This picture is, of course, based to a certain extent upon a theory of the earth's origin which seems no longer tenable; but it could probably be made to fit some of the more modern theories. If the earth was formed by the accretion of solid particles very much cooler than the sun's photosphere—though probably much hotter than the earth's surface at present—its evolution may have been radically different in some ways. Of course the temperature of the nascent earth mass may have risen considerably after it had reached something near its present size. In that case subsequent cooling could have been accompanied by changes similar to those pictured above, but it might be necessary to find some other way to account for the gaseous envelope of the earth.

Obviously we lack at the present time a unified concept of the origin of the earth compatible with all the data from astronomy, geophysics, chemistry, and geology, and in the end any successful theory must account for the evidence from all these sources. The biologist can hope to contribute little or nothing to the development of such a concept, although he needs ultimately to have such a concept if he is to formulate clearly his ideas about the origin of life and its early evolution. In the meantime he should be cautious in treating these matters in any but the most general terms, avoiding too much reliance on any specific hypothesis. One of the few generalizations he can make with assurance is that the second law of thermodynamics played a major role in determining the kind of physical environment in which life originated and particularly in determining the kinds of chemical compounds that were available for building the first living systems. This must have been the case whether the earth started as a hot gaseous mass or by the accretion of "cool" solid particles. If the biologist can help at all in the solution of the problem of the origin of the earth, it can only be by indicating what chemical compounds were essential for life's beginning there. His predictions in this regard must be based on the present constitution of living systems, and extrapolation back through the evolutionary history of these systems. Although this is not the principal object of this book, a good deal of the subsequent discussion may be pertinent to this particular problem.

Whatever the manner of formation of the earth, it seems safe to assume that explanation of the chemical complexity of its surface needs the invocation of no factors other than thoroughly established physical and chemical principles, among which the second law of thermodynamics stands out in importance, pointing the direction of each change. The dimensions are so great, however, that the interplay of physical and chemical factors might have fostered results radically different from those our common experience would lead us to predict. But the general argument may be expected to stand, even though most of the steps pictured are eventually shown to be incorrect. The surface of the earth, because of its chemical complexity, might be considered as characteristically less random—and hence to represent lower entropy—than that part of the dust cloud from which the earth formed. The formation of the earth from this apparently simpler system would, however, have been accompanied by the over-all loss of a tremendous amount of energy into space by radiation, and the total entropy change would have to be reckoned in such terms. Thus, although at first glance the apparent increase in complexity might suggest a violation of the second law of thermodynamics, further consideration reveals none.

V · LATER HISTORY OF THE EARTH

"*The best model is the one that works best. The perfect model, working infinitely well, is not for men now living.*"—REGINALD ALDWORTH DALY

IT IS generally agreed that the present earth has a core of high density, with a radius roughly one-half that of the planet itself. This core is covered by a mantle of lighter materials which displays several discontinuities. Only the outermost veneer of land surface, water, ice, and atmosphere make up the present environment of living organisms, and it is here that the interest of students of organic evolution is focused. But to gain a reasonably good view of the stage upon which life originated and the scenes among which it reached its present complexity, one needs to know something of the structure and history of the earth as a whole. About this there is diversity of opinion, and the biologist can only hope to choose a working model which may meet with a reasonable degree of acceptance by a majority of those who have made the subject their special study. It may be well for him to have some idea of the kind of evidence from which such a model must be constructed.

Core and Mantle

We owe much of our knowledge of the interior of today's earth to seismology. The differences in type and speed of waves generated by earthquake shocks—recorded by seismographs in many parts of the world—are related to the type of media through which the waves pass. Existence of a central core and several outer shells, called collectively the mantle, has been established in this way, and the different parts characterized in terms of their physical properties. Accepted estimates with regard to the core give it a radius of approximately 3,500 kilometers and a density of 10.72 times that of water. The mean radius of the earth is 6,371 kilometers, so the mantle, which may be considered to make up the remainder of the solid earth, has a thickness of approximately 2,900 kilometers. The average density of the mantle is about 2.7 near the surface, but is higher at greater depths.

Direct determinations of the chemical composition of the earth are limited to the superficial rocks, including those that have erupted from somewhat deeper parts of the mantle as igneous magmas. These are all of low density compared to the earth as a whole and cannot be expected to represent its chemical composition truly. Analysis of meteorites may give some clue to the composition of the remainder of the earth, but the validity of applying such data depends upon the nature and origin of these occasional visitors from space. It is thought that a large planet once occupied an orbit in the solar system, but broke up catastrophically. The large number of asteroids of the solar system are supposed to be the remains of that planet, as are the meteorites that from time to time reach the earth. The latter are divided into two classes according to their composition: the "iron" meteorites (they are ninety per cent iron) and the "stony" meteorites, which vary considerably in composition, but in which iron, oxygen, silicon, and magnesium predominate. By postulating that the iron meteorites represent the core, and the stony meteorites the mantle of the defunct planet, the compositions of the two types may be taken to represent those of the corresponding interior parts of the earth. In Table 2, the atomic composition of the earth has been calculated on the basis of this assumption, using the dimensions for these parts that have been given above. With regard to density and atomic composition, the igneous rocks, which come from a considerable distance below the surface, approach more closely the stony meteorites than do the superficial sedimentary rocks. This is what would be expected if the meteorites represented the constitution of the mantle as a whole, the sedimentary rocks being derived from a relatively superficial portion.

The density of solid iron at the surface of the earth is 7.76, so it must be assumed that the core with a density of nearly 11 is under tremendous pressure, the atoms being much more closely packed than in the metal we use in daily life. The temperature at the center of the earth has been variously estimated at 2,000° to 4,000°K, that is, of the same order of magnitude as that of the photosphere of the sun; but such estimates are based on extrapolation from the increase of temperature with depth within a very superficial layer, and may be far from accurate. The idea that the core is at a very high temperature remains largely as a heritage from the concept of condensation from a gas. An earth so formed should have conserved its heat after the formation of the solid crust, and might have remained at a high temperature near that of the original mass. If formed out of cooler material the internal temperature may be lower. In any event, heat released by radioactivity would have contributed to the internal temperature. Whether the core is liquid or solid is a matter of debate involving questions of the

Table 2. Quantities of the More Plentiful Elements in the Earth.

Elements		Quantities in gram atoms		
Name	Symbol	In solid earth[1] (core and mantle) $\times 10^{25}$	In hydrosphere[2] $\times 10^{22}$	In atmosphere[3] $\times 10^{20}$
Hydrogen	H	0.600	13.87	—
Helium	He	—	—	0.00002
Carbon	C	0.024	0.0003	0.0005
Nitrogen	N	—	—	2.74
Oxygen	O	10.94	6.96	0.74
Sodium	Na	0.155	0.064	—
Magnesium	Mg	1.812	0.007	—
Aluminum	Al	0.531	—	—
Silicon	Si	3.16	—	—
Phosphorus	P	0.019	—	—
Sulfur	S	0.167	0.004	—
Chlorine	Cl	0.017	0.076	—
Potassium	K	0.030	0.001	—
Calcium	Ca	0.435	0.002	—
Titanium	Ti	0.075	—	—
Chromium	Cr	0.012	—	—
Manganese	Mn	0.017	—	—
Iron	Fe	3.95	—	—
Cobalt	Co	0.203	—	—
Nickel	Ni	0.304	—	—
Copper	Cu	0.001	—	—
Zirconium	Zr	0.001	—	—
Bromine	Br	—	0.0001	—
Barium	Ba	0.001	—	—

[1] Proportion of elements based on those given by Berg, G., Chemie der Erde, in *Handbuch der Geophysik* (B. Gutenberg, ed.) (1933) *2*, 36–189. Berlin, Gebrüder Borntraeger.

[2] Based on data for the ocean, from Clarke, F. W., *The Data of Geochemistry*, 5th ed., U.S. Geol. Survey (1924), Bull. 770, Washington, G.P.O. Mass of the hydrosphere taken as 1.3×10^{24} gms.

[3] Based on data of Humphreys, W. J. *Physics of the Air*, 3rd ed., New York (1950) McGraw-Hill.

manner of transmission of earthquake shocks. It may be resistant to sudden deforming stress, but able to flow in response to continuous pressure [1]

[1] As B. Gutenberg points out, the laws describing elasticity and viscosity are based upon empirical observation under conditions obtaining at the surface of the earth—not upon physical theory. Hence they need not hold very accurately under conditions deep in the earth.

For descriptive purposes the term sima is applied to the greater part of the mantle, the term sial[2] being applied to the relatively small portion constituting principally the thin curved plates which form and underlie the continents and continental shelves. The sial is light (average specific gravity about 2.7); the sima much heavier. The former may be characterized as "granitic," the latter as "basaltic" and "peridotitic." It is thought that, in contrast to the continents, the oceans are generally floored with sima. The plates of sial which form the continental masses vary in depth with the surface structure, extending as deep as twenty-five kilometers. They may be considered as floating in the heavier underlying sima which in the case of the Pacific Ocean rises to the level of the ocean floor only a few kilometers below the highest elevation of the continents. Thus continents in sima may be compared to icebergs floating in the ocean, the greater part submerged.

How did the separate blocks of sial which constitute the continents come to be formed? Do these represent the remains of a once continuous outer sialic shell, a part of which has been lost, so that now only about thirty per cent of the earth's surface is covered by this material? Do they represent some early dissymetry of the planet, or did they result from shrinkage of the surface crust or from convection currents within a molten sphere? Have they been built by accretion of particles, or by extrusion from the interior? These are intriguing questions which cannot be definitely answered, but which could have considerable bearing on organic evolution, particularly in its later stages; for the erosion and readjustment of the continental surfaces has provided variety of environment, and by imposing factors of natural selection has encouraged multiplicity of living forms.

The greater part of the dry land of the earth is crowded into a roughly hemispherical area, the remaining hemisphere—of which the Pacific Ocean comprises a large part—being a watery one. For a long time it was thought that this arrangement of land and ocean resulted from the tearing away of about half the earth's surface, but this hypothesis has lost favor of recent years. It was supposed that the moon represents the part torn from the earth's crust as a result of tidal action within the molten core, at a time when the crust was solidified, though not to a great depth. The Pacific Ocean was supposed to be the region from which the surface layer was lost. This explanation of the origin of the moon was proposed in the last century by Sir George Darwin, and its implications as regards the earth's surface were early

[2] These words combine the first syllables of silicon and magnesium to form sima, those of silicon and aluminum to form sial (sometimes contracted to sal). The sial contains a relatively higher proportion of aluminum as compared to magnesium than does the sima.

considered by the Reverend Osmond Fisher.[3] The hypothesis was further developed by Harold Jeffreys, who now finds difficulty with the basic assumptions. Light as the moon is, a good deal of heavy material from the interior of the earth would have had to be included to account for its density—3.3 as compared to 5.5 for the earth. The association of the moon in one way or another with the apparent assymetry of the earth's surface is intriguing because so large a satellite—its mass is about one eightieth that of the earth—is unique within the solar system, and the dramatic simplicity of a catastrophic origin may have contributed to the wide acceptance of this hypothesis. Ross Gunn[4] has suggested that the moon was formed at the same time as the earth, entailing the eccentricity that resulted in the division of the surface of the earth into land and ocean "hemispheres."

The theory of continental drift, proposed early in this century, is often associated with the catastrophic origin of the moon, although this is not requisite to the theory. The crust was supposed to have broken up to form the continents which then moved apart over the molten interior; their existing outlines are such that they could be fitted together fairly neatly after the manner of a picture puzzle to form a single block covering about one-third of the earth's surface. The theory assumes that the continental masses have undergone separation and rapprochement several times during the course of evolutionary history. Different authors give somewhat different accounts based upon similarities of geological structure and biological speciation at certain periods. It is generally assumed, however, by those who support the theory that the continental mass was nearly a single unit in the late Paleozoic, and that separation into the present continents was still incomplete at the beginning of the Cenozoic Era about seventy-five million years ago.[5] Having lost favor for some years, this general idea seems again to be gaining ground, although ideas of the timing of events may have changed.

Other theories have involved changes in the earth's surface as a result of cooling of its interior. Recently, F. A. Vening-Meinesz[6] has developed the idea that the arrangement of the continents can be accounted for by a system of large convection currents set up in the early stages of a cooling earth. This theory would fix the continents in

[3] *Physics of the Earth's Crust*, New York (1889), 2nd ed., Macmillan; *Nature* (1882) *25*, 243.

[4] *J. Franklin Institute* (1936) *222*, 475–492.

[5] The names of F. B. Taylor and A. Wegener are most commonly associated with the theory of Continental Drift, but numerous others have contributed, among the most recent the late A. L. Dutoit, *Our Wandering Continents*, London (1937), Oliver and Boyd.

[6] *Quart. J. Geol. Soc. London* (1948) *103*, 191–207.

essentially their present positions very early in the earth's history, a point of view that seems quite generally accepted at present. While it is important, as regards some aspects of organic evolution, to know whether the continents assumed their present positions early or late, the details are not particularly germane to the principal thesis of this book.

The Atmosphere

In discussing the origin of the atmospheric components, the idea that the earth was formed from material of composition resembling that of the sun's photosphere will be followed generally (see Tables 1 and 2). The lightest elements must have been largely lost no matter what the manner of origin of the earth, and it is probable that some of those now present in the atmosphere have only been preserved by being combined into heavier molecules and subsequently released. With the possible exception of helium the noble gases present little problem as to their origin. They rarely combine chemically, and being gases at ordinary temperatures they would be expected to accumulate in the atmosphere except for minute quantities which might be trapped in the solidifying rocks. Helium, which is present in great quantity in the sun, has presumably, because of its low atomic weight and corresponding high atomic velocity, virtually all been lost. Neon, argon, krypton, and xenon have not yet been detected in the sun's envelope and probably are there only in very small quantities; their small proportion in the atmosphere may be associated with their relative rarity in the nascent earth.

Hydrogen, the lightest of all the elements and much the most predominant in the sun, is represented by only very small traces in the atmosphere, although a considerable amount is bound chemically elsewhere in the earth.

Nitrogen, being fairly inert chemically, is combined only in minute quantity, as referred to the total mass of the earth, but makes up the greater part of the atmosphere. Judging from Tables 1 and 2, a great deal has been lost, but one may guess that because of its combination as N_2 with molecular weight 28 or combination in other compounds from which it has been subsequently released, enough has been saved to make up four-fifths of the existing atmosphere—a very small part of the total earth, however.

Why is all the oxygen not bound as metallic oxides in the solid earth? This is the first question that is posed when one considers the origin of the molecular oxygen which makes up one-fifth of the present atmosphere. Is it possible that oxygen has existed in the atmosphere uncombined with other elements since the earth's origin as a separate body, first as atomic, later as molecular oxygen? Or was all the atmospheric oxygen at one time combined with other elements, and subsequently released? To accept either point of view requires assumptions of uncertain tenability. The origin of oxygen seems interwoven with that of water and of carbon-dioxide—all three of great biological significance—and they are best discussed together.

There is ample amount of the metallic elements in the existing earth as a whole to combine with all the oxygen that could be assumed to have existed in the nascent earth, and to assert that gaseous oxygen has always been present in the atmosphere requires the invention of some mechanism which could have prevented such combination. We might postulate some sort of spatial separation while the temperature was still so high that all the oxygen would not combine; for example, concentration of elementary iron toward the center of the earth when the temperature fell below its boiling point. Dissociation of most of the oxides of iron takes place above this temperature, so that such a separation might have removed a great deal of the iron before it could react with oxygen. Similar separation might have prevented much of the material in the mantle from combining with oxygen, but would not the remaining traces of O_2 have combined, for example, with carbon to form CO or CO_2, or with hydrogen to form HO and H_2O, all of which, being gases at the prevailing temperature, would have remained in the atmosphere? "Layering" of the gases according to their diffusion velocities, might have tended to separate some of the oxygen from other elements with which it could combine. The oxygen of the atmosphere is only a very tiny fraction of the total oxygen of the earth, making it less improbable that this small fraction could escape combination under the complex conditions that prevailed.

On the other hand, it is possible that the oxygen was virtually all combined with other elements at one time, some of it being subsequently released as O_2. It has been suggested by G. Tammann[7] that the atmospheric oxygen came from water in which it was combined. Steam at high temperature is slightly dissociated into hydrogen and oxygen, and Tammann assumed that under such conditions the hydrogen was lost into space while the oxygen, being heavier, remained. This hypothesis rests directly upon the assumption that water vapor itself could exist during the early stages of the earth's evolution, and yet not react with some of the metallic elements to form oxides more stable than water at high temperatures. Actually some spatial separation would seem to be necessary to explain the continued existence of

[7] *Zeitschrift für Phys. Chem.* (1924) *110*, 17–22.

water during a period of rapid cooling of the earth. It has been suggested by Jeffreys that water was actually held in a mixture of molten silicates and released into the atmosphere as water vapor when the silicates solidified. It seems necessary to assume under any circumstances that water has been present on the earth since early in its history.

A popular explanation of the origin of the oxygen of the atmosphere —one which has been held for many years, having probably been first suggested by Herbert Spencer[8]—is that all the O_2 now present was once combined in CO_2, or as carbonates in the solid earth, and that it was subsequently released through the activity of living organisms, that is, by photosynthesis. At the present rate of photosynthesis it would be possible to produce the total amount of oxygen now in the atmosphere in about three thousand years. A good deal of reasoning about organic evolution has been based on this idea, which seems to be losing ground, along with various other notions of terrestrial evolution, as our knowledge increases and other possibilities are envisaged. A recent approach to this problem is based on the demonstration that the O_2 from photosynthesis comes from H_2O and not from CO_2. Hence it would seem possible by comparing the isotopic composition of the oxygen in natural water and in the atmosphere to arrive at some conclusion as to whether the O_2 of the atmosphere could have arisen from photosynthesis. The results of attempts to analyze the situation seem, however, to be somewhat conflicting and equivocal. [9]

Malcolm Dole has proposed that O_2 is being formed at present in the stratosphere, by photochemical and ionization processes, from other molecules such as oxides of carbon and nitrogen. Such photochemical reactions are prevented at lower altitudes because the shorter ultraviolet wavelengths are removed by a blanket of ozone, itself formed by a photochemical reaction. Dole's proposal offers a plausible mechanism for the origin of atmospheric oxygen without the participation of living organisms.

Possibly the oxygen of the atmosphere originated from more than one source, perhaps including all those mentioned and others as well; a quantitative evaluation of the contributions of various mechanisms seems impossible considering our uncertainty as to the mode of origin of the earth. Recently, H. L. Urey has brought strong support to the concept of a primitive reducing atmosphere, from which O_2 was absent. This has been proposed earlier by A. I. Oparin (1938, 1953), who based on it his ideas of the anaerobic origin of life. The bearing

[8] *Philosophical Magazine* (1844), *24*, 3rd series, 90–94, with a skeptical note appended by the editor, amusing now when one considers the subsequent degree of acceptance of this idea.

[9] E.g. see Kamen, M. D., and Barker, H. A., *Proc. Nat. Acad. Sci.* (*U.S.*) (1945) *37*, 8–15; Dole, M., *Science* (1949) *109*, 77–81.

of the primitive atmosphere on the problem of life's origin will be discussed in Chapter X.

Carbon dioxide is an all important constituent of the atmosphere as regards living systems as a whole, although it is present in very small amount. As would be expected in the great excess of O_2, all the carbon of the existing atmosphere is found as CO_2. But the origin of CO_2 from a strictly reducing atmosphere may present problems. Volcanic origin has been suggested (see W. W. Rubey, 1951). Anearobic metabolism could conceivably have contributed.

Because of their similar masses and distances from the sun, the planets Venus and Mars might be expected to have atmospheres somewhat similar to that of the earth. This is not the case, however, and this does not diminish the difficulty of explaining the presence of molecular oxygen, water, and carbon dioxide on the earth. Spectroscopic exploration has placed the highest concentration of O_2 on both of these neighboring planets at less than one one-hundredth part of that in the earth's atmosphere, and even lower limits have been set for water vapor. On Mars there may be considerable water in the form of ice, although liquid water may be absent. Most puzzling perhaps is the fact that Venus has over one hundred times as much CO_2 as the earth; whereas the concentration of this gas on Mars seems relatively low, although perhaps greater than in the earth's atmosphere. This has been interpreted as indicating the absence of photosynthesis on Venus and its possible presence on Mars. When one considers some of the difficulties of the origin and rather special character of green plant photosynthesis this argument may not appear too cogent.

A small quantity of ozone exists in the contemporary atmosphere, largely restricted to the stratosphere, where if it were all concentrated as pure gas it would form a layer only about three millimeters thick. Yet this layer is of greatest importance to living organisms, for it absorbs a considerable part of the ultraviolet radiation of sunlight that would otherwise reach the earth's surface, thus filtering out rays which are generally destructive to living organisms. Ozone, O_3, is formed by the action of short ultraviolet radiation on O_2 or other oxygen-containing molecules in the stratosphere, while the opposite effect is brought about by the somewhat longer wavelength ultraviolet which the ozone itself absorbs. Thus, there is a more or less constant balance between formation and dissociation of this substance. If at one time there was no molecular oxygen in the atmosphere of the earth, there was likewise no ozone, and the earth's surface would at that time have been exposed to the action of a great amount of short wavelength ultraviolet radiation which does not now reach it. Such radiation could have brought about photochemical reactions which do not occur there today. This has been the basis of certain speculations regarding the origin of life. If the atmospheric oxygen itself originated from photochemical reactions, we see here an involved series of interrelated processes which may have contributed importantly to the complex matrix from which life originated.

The Hydrosphere

Whatever the manner of formation of the earth, its surface probably enjoyed at some point in its evolution a temperature higher than the boiling point of water, and it is reasonable to assume that the surface waters arose principally from condensation of water vapor from the atmosphere. At 100°C, when condensation could begin, the crust of the earth would of course have been solid. Water vapor is a stable compound which dissociates only at very high temperatures, and its molecular weight is high enough so that it would not readily escape.[10] The apparent absence of water on Venus and Mars, in quantity comparable to that on the earth, is all the more puzzling for this reason.

Attempts to make direct estimates of the age of the ocean on the basis of its salt content meet with difficulty. Those based on the amount of sodium in the sea and the present rate of erosion put the age at only about fifty million years, a figure that was once accepted as the age of the earth. This figure is only a fraction of that now attributed to the oldest sedimentary rocks, the formation of which depended upon the existence of oceans and continents.

The Continental Surfaces

Differentiation of the sial from the sima probably took place after the earth's surface was essentially solid. Extrusion of molten magmas from the deep interior, and accretion of eroded material, have both been major factors in building the continents, but opinion differs as to their relative importance in the initial stages. During at least the last two billion years the sculpturing of the continental surfaces has resulted principally from a succession of elevations followed by periods of erosion. The eroded material formed deposits on the submerged areas of the continents, and where these have been subse-

[10] Dauvillier A. and Desguin E. (*La genèse de la vie*, Paris (1942) Hermann) propose that the water vapor that condensed to the hydrosphere was formed by the reduction by hydrogen of ferric to ferrous iron.

quently elevated we now find sedimentary rocks covering large areas, their layered character revealing the manner of formation. Molten magmas rising through cracks in the surface have in many places transformed these sedimentary rocks. The result is a complex surface not always easy to interpret.

There has been a continual process of adjustment of the elevation of surface units. The masses of the mountains are not to be regarded as merely piled up on the surface, but as extending much deeper into the earth than they rise above. As the weight of the exposed part of a mountain is decreased by erosion it tends to rise again, being pushed up by the readjustment of the masses. At the same time the weight in the surrounding areas is increased by the addition of the eroded material, tending also to cause the mountain to rise. This adjustment and readjustment to reach a balance is known as *isostasy.* For the large-scale units of the earth's crust the balance seems nearly complete at present, although this is not true for smaller units.

Erosion has resulted principally through the action of water, attrition by suspended particles being a major factor. In addition periodically appearing ice masses have scoured large areas of the earth and have caused isostatic adjustments as they formed and regressed. Winds have carried fine-grained solid material, depositing it in certain areas. Living organisms have brought about the deposition of large quantities of calcareous and siliceous material in the ocean bottoms and reefs, and of bitumen in land areas. All these factors have brought continuous and periodic alterations of the terrestrial landscape which have had important effects on the evolution of living organisms.

The Role of the Sun

About 1.3×10^{21} kilogram-calories of solar energy strikes the profile of the earth each year[11] –in the course of three billion years about 3.9×10^{30} kilogram-calories, assuming that the temperature of the sun has remained approximately constant during that time. This stupendous quantity of energy maintains the temperature of the earth's surface, feeds all living things thereon, and provides the energy for the rains, winds, and ocean currents—all factors that have helped in different ways to create the varied aspect of the earth we know.

Some of the solar radiation is reflected back from the earth, the amount depending upon the nature of the surface—rock, water, ice, green foliage, clouds, and so forth. The remainder of the incident radiation is absorbed and tends to raise the temperature of the surface either directly or indirectly. If the temperature of the surface (at

[11] Based on the solar constant, 0.0019 kilogram-calories per square cm. per minute, the estimated amount of solar energy reaching the outer atmosphere.

present about 10° to 15°C) is to remain approximately constant, the earth must somehow get rid of a quantity of energy equal to that gained by the absorption of sunlight plus a very small amount (negligible in the present argument) gained by conduction from the hot interior of the earth, from radioactivity in the crust, and from still lesser sources. The loss is actually accomplished by radiation from the earth itself into space.

An appreciation of the balance between energy gain and energy loss, which determines the earth's temperature, requires an understanding of the spectral distribution of the incoming and outgoing radiation. Curve 0 of Figure 4 represents the spectral distribution of sunlight outside the earth's atmosphere as estimated from measurements in which correction is made for absorption by the atmosphere. The maximum of the curve is at about 0.48 μ. Curves 1 and 2 describe the spectral distribution of sunlight after passing through the atmosphere; their notched appearance is due to absorption by water vapor. The limitation of the ultraviolet, at the short wavelength end, is due to ozone. The earth at a temperature of approximately 287°K should radiate energy with the spectral distribution corresponding to that of a black body at that temperature, which is represented in Figure 5, the maximum of emission lying at about 10 μ.

The actual temperature at the bottom of the atmosphere, where living organisms reside, is dependent upon what is commonly called the *greenhouse effect*. In constructing houses for growing green plants, advantage is taken of the fact that ordinary window glass transmits most of the energy of sunlight, but does not transmit the greater part of the black body radiation from the interior. The radiant energy of sunlight enters the greenhouse, and is largely absorbed there by the plants and other surroundings. These objects re-emit the absorbed energy as radiation of much longer wavelengths, with a spectral distribution about that represented by the black body curve in Figure 5. This long wavelength radiation cannot pass out through the glass, so tends to be trapped inside the greenhouse, and hence to raise the temperature of the interior. All bodies of matter emit radiant energy, and the temperature of any given body depends on the exchange of radiant energy with the other bodies which make up its surroundings. The relationships may be quite complex, and the balance sheet difficult to draw up. For example, although the glass roof of the greenhouse transmits most of the sunlight that strikes it, some is absorbed and this portion serves to raise the temperature of the glass above that of its surroundings. The glass emits this absorbed energy as black body radiation some of it going to the outside, and some of it into the greenhouse. Some of the heat absorbed by the glass is also distributed to the interior of the greenhouse

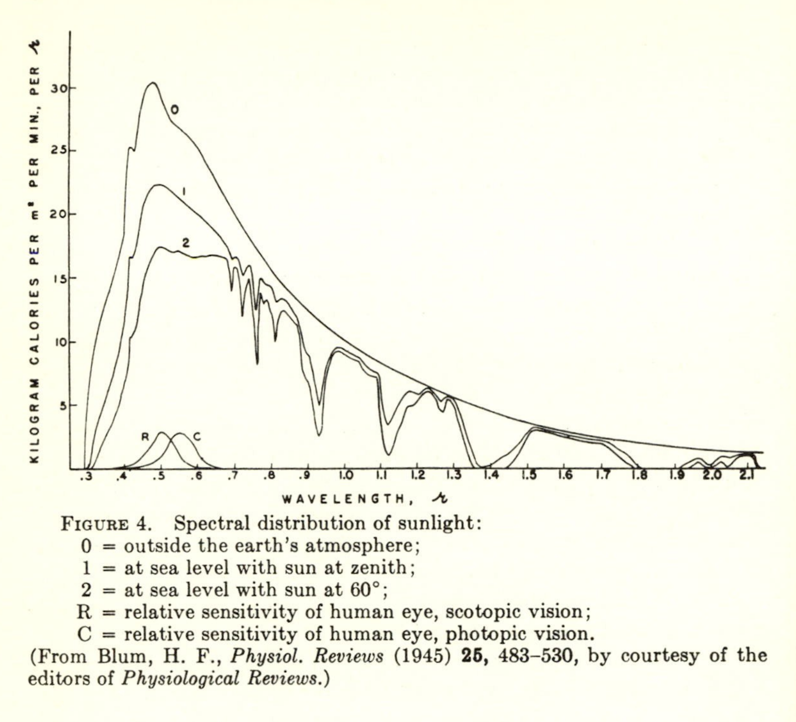

FIGURE 4. Spectral distribution of sunlight:
0 = outside the earth's atmosphere;
1 = at sea level with sun at zenith;
2 = at sea level with sun at 60°;
R = relative sensitivity of human eye, scotopic vision;
C = relative sensitivity of human eye, photopic vision.
(From Blum, H. F., *Physiol. Reviews* (1945) **25**, 483–530, by courtesy of the editors of *Physiological Reviews.*)

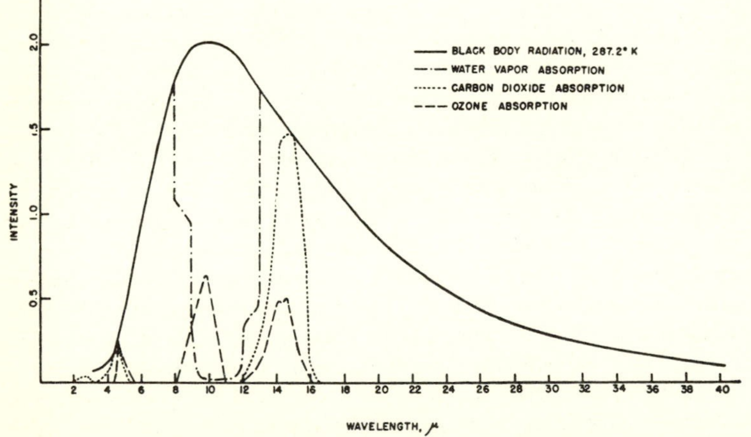

FIGURE 5. Spectral distribution of radiation from the earth (black body at 287°K), and relative absorption by atmospheric gases. Note transparency of water vapor in the region of 10 μ. (Redrawn from Humphries, W. J., *Physics of the Air.* New York (1929) McGraw-Hill.)

by conduction and convection, but for our simple analogy this will be neglected.

The atmosphere of the earth exerts much the same effect as the glass of the greenhouse, allowing the greater part of the incident solar energy to pass in, but absorbing a considerable proportion of the earth's own radiation. The principal atmospheric absorber for the entrant sunlight is water vapor, absorption by ozone being a minor factor quantitatively; the other gases are virtually transparent. Absorption of the outgoing radiation from the earth is again largely due to water vapor, with CO_2 and ozone playing lesser roles. The absorption spectra of the absorbing gases relative to the emission of a black body at the temperature of the earth are shown in Figure 5. As indicated there, water vapor strongly absorbs a large proportion of the wavelengths emitted by the earth, but has a distinct minimum of absorption at about 10 μ corresponding quite closely to the maximum of the earth's emission. Thus water vapor absorbs a considerable part of the radiation emitted by the earth and hence contributes greatly to the greenhouse effect, but at the same time allows some of this radiation to escape. The part absorbed tends to warm the atmosphere, and just as the warm glass of the greenhouse tends to raise the temperature of the interior, the water vapor tends to raise that of the earth's surface below it. This surface, or any object on it, is constantly exchanging radiation with the water vapor in the atmosphere, so the temperature of the surface is closely dependent upon the amount and temperature of this vapor. Were it not for the fact that water vapor transmits a good part of the radiation emitted by the earth due to its transparency in the region of 10 μ, the surface temperature would be considerably higher.

In the infrared region of the spectrum, ozone is a much less important absorber than water vapor, but it must exert a similar effect; its quantity varies periodically with the season and may account for some fluctuations of temperature. Carbon dioxide is present in such small quantity at present that it can have little effect on temperature; if it was present long ago in much greater quantity but has been reduced by photosynthesis, as is often assumed, it could have influenced the early climate of the earth.

If it were not for the rotation of the earth, the surface would reach very high temperatures on the side toward the sun. By rotation the warm earth surface tends to dissipate at night the heat accumulated during the day, the atmosphere considerably moderating the diurnal extremes.

The average radiation from the sun, compared to which the other factors above are relatively small, has probably not changed greatly during the last 3.8 billion years, which is the age now assigned to the oldest sedimentary rocks. It may be assumed that the temperature of the surface of the earth has remained relatively constant during the greater part of that time and perhaps much longer. The average temperature should have been somewhat higher, if anything, in the distant past. There was once more heat from radioactivity of elements such as radioactive potassium, present in the nascent earth but long decayed. High concentration of carbon dioxide may have interfered with heat loss and kept the temperature appreciably higher for a time. Thus, while it is quite certain that the temperature was at one time above that compatible with life, there is good reason to think that for the last three billion years and perhaps longer the surface of the earth has enjoyed a temperature appropriate for the evolution of living systems. There have doubtless been considerable fluctuations of temperature involving changes of a few degrees—minor from the standpoint of the earth as a whole, but major changes for living organisms—as the paleontological record indicates. This record, and that of geology, tell us clearly that at certain times great areas of the earth, which now enjoy temperate or even tropical climates, were covered with ice. They do not tell us with certainty, however, whether these ice ages were due to general or only local changes in temperature. It has been assumed by some that the evidence of ice where it is not found today indicates that changes in the position of the poles of the earth have occurred from time to time. Such shifts need not have involved changes in the axis of rotation of the earth, but could result from a slipping of the surface layers, thus changing the positions of the poles with respect to the continents.

Most other hypotheses regarding the ice ages involve a general change in the earth's surface temperature resulting in an expansion of the area of polar ice. Such lowering of temperature has been credited to changes in the amount of solar energy reaching the earth, to change in the magnitude of the greenhouse effect caused by changes in atmospheric CO_2, and to periodic minor changes of the axis of rotation with respect to the sun. An interesting hypothesis formulated by the late Gilbert N. Lewis [12] may be mentioned because it involves a factor that is sometimes neglected. Reflection of solar energy is higher from ice than from most other terrain, and this tends to decrease the temperature of the air over glacier covered areas so that water vapor condenses there. If there were, for some reason or other, a general increase in precipitation there would be a tendency to remove water from the air by freezing it in the glacier covered areas, and thus "siphoning" it off from other areas. This could result in a self-accelerated runaway expansion of the glaciers, which would only be suppressed when the

[12] *Science* (1946) *104*, 43–47.

temperature, particularly in the tropics, was reduced sufficiently to lower evaporation and thus limit the siphoning off of water into the area of the glaciers. The latter effect would tend to reverse the process, and cause the glaciers to regress. It is possible that numerous factors were concerned in the advance and recession of the ice; perhaps all those mentioned and others in addition have played roles of varying importance.

The winds and ocean currents result from temperature differences at different parts of the earth, and hence their energy is derived ultimately from the sun. The rains represent masses of water raised into the air, which then precipitate on elevated parts of the land, and supply water power for erosion. This energy, of course, comes also from the sun, as does the supply of energy that maintains living organisms, the energetics of which will be discussed later on. Thus, in any attempt to consider the evolution of the earth as that of a thermodynamic system, it should not be forgotten that the complexity of its surface derives in great part from the energy supplied by the sun, and that the earth and the sun must be regarded as a coupled system.

In considering the evolution of the nonliving world, one cannot help but be impressed by the universal character of many of the contributing factors, for example the nature of the chemical elements, their distribution, and the thermodynamic properties which govern their combinations and determine the kind of molecules that will be formed from them under given circumstances. On the other hand one is also impressed by the apparently unique and "accidental" character of many of the factors which make the earth's surface and the environment of living organisms what they are. In the next chapter and in subsequent ones the importance of various factors, both universal and accidental, will be weighed with regard to their bearing on the evolution of living systems.

VI · THE FITNESS OF THE ENVIRONMENT

"*A quoi tient la destinée. Si le silicium avait été un gaz, je serait major general.*"—*attributed to* JAMES ABBOT MCNIELL WHISTLER.

IN 1913 there appeared a small volume entitled *The Fitness of the Environment* written by Lawrence J. Henderson, which biologists and others received with varied response. Many found in this book a new and stimulating idea; others saw only platitudes. Some were frankly puzzled by it, and it has even been suggested that it was written with tongue in cheek. The opening words of Henderson's preface express his thesis briefly. "Darwinian fitness is compounded of a mutual relationship between the organism and the environment. Of this, fitness of the environment is quite as essential a component as the fitness which arises in the process of organic evolution; and in fundamental characteristics the actual environment is the fittest possible abode of life." He goes on to support his idea with evidence that certain aspects of environment render it peculiarly advantageous for the abode and development of living organisms.

There have been so many conflicting opinions regarding Henderson's book that perhaps a brief attempt should be made to forestall certain misapprehensions. As he freely admits, he was trying to formulate in more acceptable terms a concept that had been expressed before, particularly in the pre-Darwinian period, but which had lacked strictly physical explanation. In attempting to find order in a complex mass of fact, he was obviously groping, at times, and this makes *The Fitness of the Environment* difficult to evaluate at some points, allowing the reader latitude in the ideas he ascribes to the author—Henderson has been accused of both mysticism and over-indulged materialism. But after careful rereading of doubtful passages one is usually impressed with the clarity of thought as well as the grasp of evidence within his generalization. Some of Henderson's interpretations based on physical chemistry must be revised because that science has advanced apace since the book was written, and some modern readers particularly concerned with details may miss the main current of his argument on that account.

Much the same thing could be said about Darwin's *Origin of Species*, however, yet few of us allow this to detract from the importance of that author's great contribution to evolutionary theory, the concept of origin of species by variation and natural selection. Henderson's idea may not compare in importance with Darwin's—though I should not wish to defend this position—but to me it seems that the fitness of the environment has entered into evolutionary thinking to a much less extent than it might.

Throughout, it must be understood that Henderson is not trying to substitute something in place of evolution by natural selection. He repeatedly asserts his acceptance of Darwinism, as in the following passage: "Today, after a half century there is no longer room for doubt that the fitness of organic beings for their life in the world has been won in whole or in part by an almost infinite series of adaptations of life to its environment, whereby, through a corresponding series of transformations, present complexity has grown out of former simplicity." What Henderson attempts to demonstrate is a reciprocal relationship.

In illustrating fitness of the environment, Henderson sometimes deals with properties of components of living organisms as they refer directly to the organisms themselves, rather than to the surroundings of the organism. That is, not only does he reason in terms of the materials external to living organisms, and in terms of the *milieu interieur*[1] in the restricted sense of an internal environment provided by the body fluids; but inclusively, in terms of the intimate components of living systems themselves. This seems a necessary part of his general thesis. Once accepted, it is impossible to treat the environment as a separable aspect of the problem of organic evolution; it becomes an integral part thereof.

A detailed defense of all of Henderson's arguments does not seem in order, and of course the interpretation of fitness which is developed in this chapter should stand on its own merits without the support of authority. Thus I must take full responsibility for the ideas which appear here, but in doing so I must also pay highest tribute to the source from which they derive. In order to illustrate the use of the term "fitness," I shall need examples; and in choosing these I shall make additions to as well as omissions from Henderson's material. I, too, shall restrict my discussion principally to a few compounds of the elements hydrogen, oxygen, and carbon, and, as a starting point, to that ubiquitous substance, water.

[1] This term was introduced by Claude Bernard (*Introduction à l'étude de la Medicine Experimentale*, 1865, J. B. Baillière et Fils.) to describe the stabilized internal conditions provided in higher organisms, for example, by the blood stream, so that the living systems themselves are effectively in contact with an internal environment rather than the external one.

The Fitness of Water

Water makes up perhaps eighty to ninety per cent of all living organisms, and may be regarded as their principal environmental component, since even forms living in air maintain an aqueous internal environment in one way or another. Most of the water on the earth is in the liquid state, but it is also of importance as an environmental factor when in the vapor state, and even as a solid.

Water seems admirably fitted for the major role it plays in maintaining a relatively constant temperature for the earth's surface, a matter of paramount importance to living organisms, which can survive only within a very restricted range of temperature. It owes this aspect of its fitness to several properties.

The *heat capacity*[2] of liquid water is among the greatest for compounds of its type in the liquid state. This means that compared with most other substances, a large quantity of heat is required to cause a given increase in the temperature of a given quantity of water. Hence this substance serves effectively to maintain temperature at a constant level, as its use in the laboratory in constant temperature baths testifies. The quantity of water on the earth's surface—about 1.25×10^{24} grams, or enough to form a layer 2.5 kilometers deep if spread evenly—thus tends to prevent sudden rises of temperature, as for example between night and day. Not only the temperature of the ocean itself is controlled in this way, but to a considerable extent that of the atmosphere in its immediate vicinity. The specific heat of air is less than that of water, so the atmosphere is subject to much more sudden changes in temperature than is the ocean; but the atmospheric water vapor tends to maintain equilibrium with the ocean, and hence has a stabilizing effect. That is why climates in the neighborhood of large bodies of water are the most equable. The heat capacity of the sea, by buffering sudden changes in local temperature, prevents the occurrence of catastrophic ocean currents and winds, which would result from rapid changes. As the preponderant component of living systems, water serves also to prevent too sudden changes in the temperature of these systems themselves; this may sometimes be an important factor in the survival of forms that inhabit the land.

Another property of water which assists in the control of temperature is its high *latent heat of vaporization.* In order to change one gram of liquid water into water vapor it is necessary to put in between five and

[2] We are concerned here with the heat capacity at constant volume which is commonly called the *specific heat.* Liquid water is the substance used as the basis of calorimetry and its specific heat is taken as unity for this purpose. One gram-calorie is defined as the quantity of heat required to raise one gram of water from 14.5°C to 15.5°C.

six hundred calories of heat (depending upon the temperature), which means that this much energy must be withdrawn from the immediate environment. The condensation of the same quantity of water requires that an equal amount of heat be transferred from the water to the environment. The air above a body of water can take up a given amount of water vapor, the amount increasing with the temperature; so, as the temperature rises, the amount of water vapor in the atmosphere tends to increase. The evaporation of water entails the input of a large amount of heat because of its high latent heat of vaporization, and hence tends to prevent a rapid rise in ambient temperature when there is a sudden input of heat; this occurs for example at sunrise.

The latent heat of vaporization is an important factor in regulating not only the earth's temperature, but also the temperatures of land plants and animals. Were it not for the great amount of heat dissipated by land plants due to evaporation through transpiration, these organisms could not survive the temperatures and exposures they frequently encounter. Evaporation is likewise an important factor in preventing undue increase in temperature in those land animals which are able to evaporate water by sweating or other means. For example, a man exposed to sunlight in a hot desert climate, where the ambient temperature is higher than that of his body, must depend largely upon evaporation for cooling; he may under such conditions evaporate as much as one liter of water per hour. A less important channel of heat loss under such conditions may be radiation to the cool atmosphere at higher levels, principally to the water vapor in that atmosphere.[3]

Water also possesses a high *latent heat of fusion.* This, too, assists in the regulation of the temperature of the ocean and, hence, of the earth's surface in general. To change one gram of liquid water to ice at the freezing point it is necessary to remove eighty calories from the water, and this of course tends to warm the environment. Thus, if the temperature of water falls to the freezing point, it cannot go below this temperature without losing a considerable amount of heat to the environment, and this tends to prevent the temperature of large bodies of water from falling much below this value.[4] When ice melts the same amount of heat must be given up by the environment.

An anomalous property of water, namely that its greatest density is reached at 4°C instead of at the freezing point, may, as Henderson pointed out, have environmental importance. This property tends to prevent freezing from the bottom up, since water in the neighborhood of 4°C is heavier than water at the freezing point and will sink, tending

[3] Blum, H. F., *J. Clinical Invest.* (1945) *24*, 712–721.

[4] Due to the presence of high concentration of salts, the freezing point of sea water is depressed a few degrees.

to displace warmer water at the bottom. Thus freezing begins at the surface, and the bottom freezes last, and so organisms living at the bottoms of fresh water ponds and lakes are protected. Organisms in the primitive ocean, if they existed there before it became highly saline as the result of erosion of the land surfaces, may have enjoyed protection for the same reason; but the situation there is now somewhat different, the same effect being accomplished in a different way. The ice that freezes out is less dense than the salt water itself and therefore floats on the surface. The freezing out of pure water concentrates the sea water in the neighborhood of the ice and increases the density of the liquid so that it tends to sink. Thus, the same thing is accomplished as in fresh water.

I have mentioned already that water vapor is formed in the tropics as a result of the high temperatures prevailing there, and in the cooler temperate and polar regions water condenses and may become ice. This entails a considerable exchange of heat. Because of the high latent heat of vaporization, a good deal of heat is removed from the tropical environment in the formation of water vapor from liquid water and is given up in the cooler latitudes when the vapor condenses. If snow or ice forms, still more heat is released to the environment, corresponding to the latent heat of fusion of water. Most of this heat is eventually returned to lower latitudes by the melting of the ice and by ocean currents, the result being a more uniform temperature over the surface of the earth. G. N. Lewis's suggestion that the successive ice ages of geologic history each represented a runaway of this process was mentioned in Chapter V.

Water in the vapor state plays a role of profound importance not only in maintaining constancy of temperature but in determining its absolute value. The "greenhouse" effect has already been discussed in the last chapter, where it was pointed out that water vapor in the atmosphere permits most of the sunlight to pass through freely, but absorbs to a considerable extent the infrared radiation re-emitted by the earth itself. The alternation of day and night due to the earth's rotation allows the heat energy accumulated at one time to be lost at another, so that a relative balance is achieved. A special spectral characteristic of water vapor, i.e. a zone of transparency just at the maximum of emission for the earth, is illustrated in Figure 5. Slightly different spectral characteristics or a somewhat different latent heat of vaporization might have resulted in a quite different temperature for the earth's surface. A somewhat different mass or slower rotation could have given the earth a different temperature, perhaps one quite incompatible with life as we know it. The conditions on our two nearest planet neighbors, Venus and Mars, exemplify for us temperature environments that might have prevailed on the earth under conditions not too different from those that exist.

A few other properties of water may be briefly mentioned. Its character as a "universal solvent" provides for the ready transport of a multitude of substances throughout living systems, and their exchange between these systems and the environment. The ability of water to produce ionization of substances in solution, its high surface tension, and the ability of its molecules to associate with each other or with other particular species of molecules are factors whose environmental fitness is more difficult to assess. The list of properties contributing to the fitness of water could be still further extended, but without seeking more examples let us consider what determines those properties of this substance, that underlie its fitness. Are these properties unique to this substance and, if so, to what do they owe their uniqueness? Before an attempt can be made to answer these questions, it will be necessary to inquire briefly into factors which determine the properties of atoms and molecules in general.

The Nature of Atoms and Molecules

Any atom is composed of a central nucleus made up of particles of mass about equal to that of the hydrogen atom, which is taken as the *unit of mass;* these are *protons*, which have unit positive electrical charge, and *neutrons*, which have no electrical charge. The charge on the nucleus is equal to the number of protons and is known as the *atomic number;* it is this number which characterizes the atom *chemically*, all atoms of the same atomic number having the same chemical properties. The elements are arranged in the periodic table, in the order of their atomic numbers, beginning with hydrogen 1, and ending with californium 98, as shown in Figure 7.

The atomic weights of the elements depend upon the number of particles of unit mass of which they are composed. The same number of units of mass in the nucleus is not always accompanied by the same charge on the nucleus, however, so that atoms with the same atomic number—and, hence, the same chemical properties—may have different atomic weights. These are *isotopes;* for example, there are five isotopes of carbon, which are symbolized as ${}_6C^{10}$, ${}_6C^{11}$, ${}_6C^{12}$, ${}_6C^{13}$, and ${}_6C^{14}$. The subscript indicates the atomic number; the superscript the mass number, which is virtually equal to the atomic weight. The atomic number is the same for all these isotopes, indicating that the chemical properties are identical. The mean atomic weight is the chemical atomic weight which is given for most of the known elements in Table 3. In subsequent discussion we will be principally concerned with the chemical properties of the elements, which are determined by their atomic numbers.

Table 3. International atomic weights, 1949. Reproduced by courtesy of the editors of the *Journal of the American Chemical Society.*

	Symbol	Atomic Number	Atomic Weight[a]		Symbol	Atomic Number	Atomic Weight[a]
Actinium	Ac	89	227	Neptunium	Np	93	[237]
Aluminum	Al	13	26.97	Neon	Ne	10	20.183
Americium	Am	95	[241]	Nickel	Ni	28	58.69
Antimony	Sb	51	121.76	Niobium			
Argon	A	18	39.944	(Columbium)	Nb	41	92.91
Arsenic	As	33	74.91	Nitrogen	N	7	14.008
Astatine	At	85	[210]	Osmium	Os	76	190.2
Barium	Ba	56	137.36	Oxygen	O	8	16.0000
Beryllium	Be	4	9.013	Palladium	Pd	46	106.7
Bismuth	Bi	83	209.00	Phosphorus	P	15	30.98
Boron	B	5	10.82	Platinum	Pt	78	195.23
Bromine	Br	35	79.916	Plutonium	Pu	94	[239]
Cadmium	Cd	48	112.41	Polonium	Po	84	210
Calcium	Ca	20	40.08	Potassium	K	19	39.096
Carbon	C	6	12.010	Praseodymium	Pr	59	140.92
Cerium	Ce	58	140.13	Promethium	Pm	61	[147]
Cesium	Cs	55	132.91	Protactinium	Pa	91	231
Chlorine	Cl	17	35.457	Radium	Ra	88	226.05
Chromium	Cr	24	52.01	Radon	Rn	86	222
Cobalt	Co	27	58.94	Rhenium	Re	75	186.31
Copper	Cu	29	63.54	Rhodium	Rh	45	102.91
Curium	Cm	96	[242]	Rubidium	Rb	37	85.48
Dysprosium	Dy	66	162.46	Ruthenium	Ru	44	101.7
Erbium	Er	68	167.2	Samarium	Sm	62	150.43
Europium	Eu	63	152.0	Scandium	Sc	21	45.10
Fluorine	F	9	19.00	Selenium	Se	34	78.96
Francium	Fr	87	[223]	Silicon	Si	14	28.06
Gadolinium	Gd	64	156.9	Silver	Ag	47	107.880
Gallium	Ga	31	69.72	Sodium	Na	11	22.997
Germanium	Ge	32	72.60	Strontium	Sr	38	87.63
Gold	Au	79	197.2	Sulfur	S	16	32.066
Hafnium	Hf	72	178.6	Tantalum	Ta	73	180.88
Helium	He	2	4.003	Technetium	Tc	43	[99]
Holmium	Ho	67	164.94	Tellurium	Te	52	127.61
Hydrogen	H	1	1.0080	Terbium	Tb	65	159.2
Indium	In	49	114.76	Thallium	Tl	81	204.39
Iodine	I	53	126.92	Thorium	Th	90	232.12
Iridium	Ir	77	193.1	Thulium	Tm	69	169.4
Iron	Fe	26	55.85	Tin	Sn	50	118.70
Krypton	Kr	36	83.7	Titanium	Ti	22	47.90
Lanthanum	La	57	138.92	Uranium	U	92	238.07
Lead	Pb	82	207.21	Vanadium	V	23	50.95
Lithium	Li	3	6.940	Wolfram			
Lutetium	Lu	71	174.99	(Tungsten)	W	74	183.92
Magnesium	Mg	12	24.32	Xenon	Xe	54	131.3
Manganese	Mn	25	54.93	Ytterbium	Yb	70	173.04
Mercury	Hg	80	200.61	Yttrium	Y	39	88.92
Molybdenum	Mo	42	95.95	Zinc	Zn	30	65.38
Neodymium	Nd	60	144.27	Zirconium	Zr	40	91.22

A value given in brackets denotes the mass number of the most stable known isotope.

One of the great generalizations of chemistry is the periodic arrangement of the elements; that is, when the elements are arranged in order of atomic number, certain properties recur in regular sequence. An example is the periodic variation of the property, atomic volume, illustrated in Figure 6, which has been reproduced from *The Fitness of the*

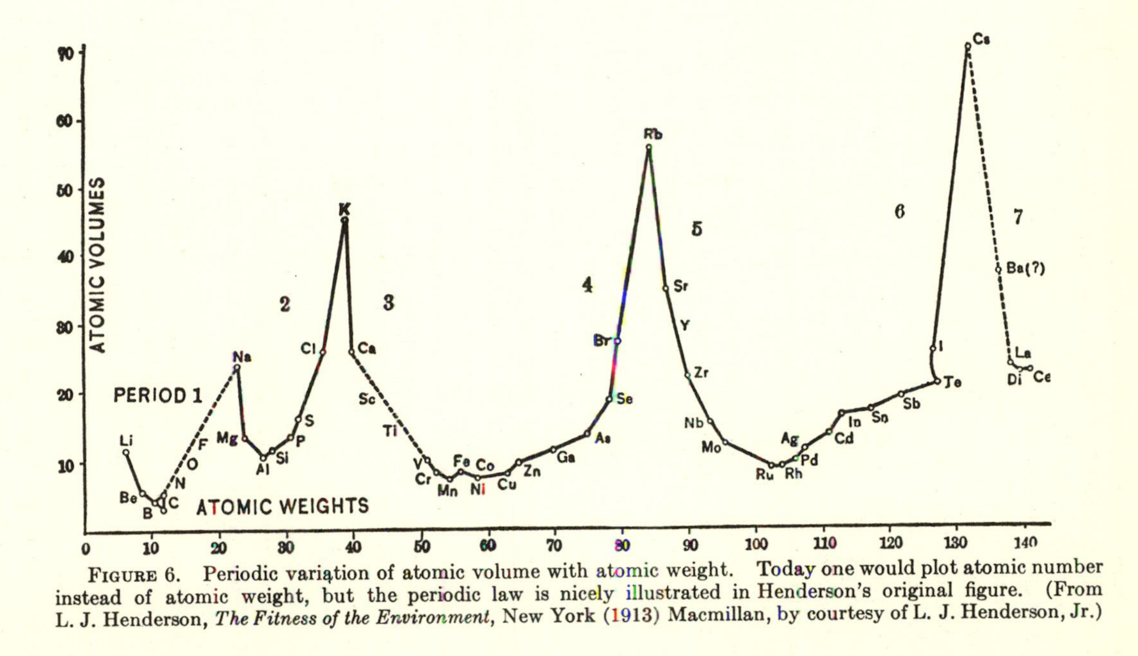

FIGURE 6. Periodic variation of atomic volume with atomic weight. Today one would plot atomic number instead of atomic weight, but the periodic law is nicely illustrated in Henderson's original figure. (From L. J. Henderson, *The Fitness of the Environment*, New York (1913) Macmillan, by courtesy of L. J. Henderson, Jr.)

Environment. In this diagram, atomic volume is plotted against atomic weight, the concept of atomic number just being developed by Moseley at the time that book was written. Today one would plot atomic number rather than atomic weight and could include another period of elements then unknown for the most part.

In the periodic table, shown in Figure 7, the elements are arranged in seven periods (vertical columns I to VII); the first consisting of only two elements, the next two of eight, followed by two periods of 18, and these by one of 32 and an apparently incomplete one of 12. Elements enjoying comparable position in these periods constitute in each case a group having comparable properties. For example, the element of highest atomic number in each of the periods, He, Ne, A, Kr, Xe, Rn, constitute the group of noble elements; each is an inert gas whose atoms rarely combine to form molecules. After the first period, the element having the atomic number just below that of the noble element in each of the periods is a halogen, F, Cl, Br, I, At; and the elements of the group having the lowest atomic number in each period are the alkali metals, Li, Na, K, Rb, Cs, and Fr. The elements of each of these groups exhibit common, well-known properties which distinguish it as a group from other elements. At the time *The Fitness of the Environment* was written, little was known of the underlying reasons for such an arrangement, and Henderson could only follow rather vague general relationships in trying to judge the relative fitness of elements and the molecules formed from them. The factors determining the periodic arrangement are much better understood today, and this added knowledge helps to support his basic concept.

The normal atom of any element has surrounding the nucleus a number of *electrons*—particles of negligible weight with unit negative charge—equal to the number of positive charges on the nucleus, i.e. the atomic number. The nucleus makes up a small part of the volume of the atom and the outer electrons a much tinier part, but the latter are in such rapid motion that in any finite time they virtually fill all the volume other than that of the nucleus. For purposes of visualization the volume occupied by the electrons may be thought of as "mushy" in character. The paths of the electrons are not strictly determinable, but they are restrained to move within certain limits, and as a result tend to form a series of shells around the nucleus, there being room for only a limited number of electrons in each shell. The periodic arrangement of the elements follows the order of these shells, each period corresponding to a shell. Proceeding through the table in order of atomic number, the inner shells are filled first, so that if there are not enough positive charges on the nucleus to hold electrons to fill all the shells, it is the outer shell that is left incomplete.

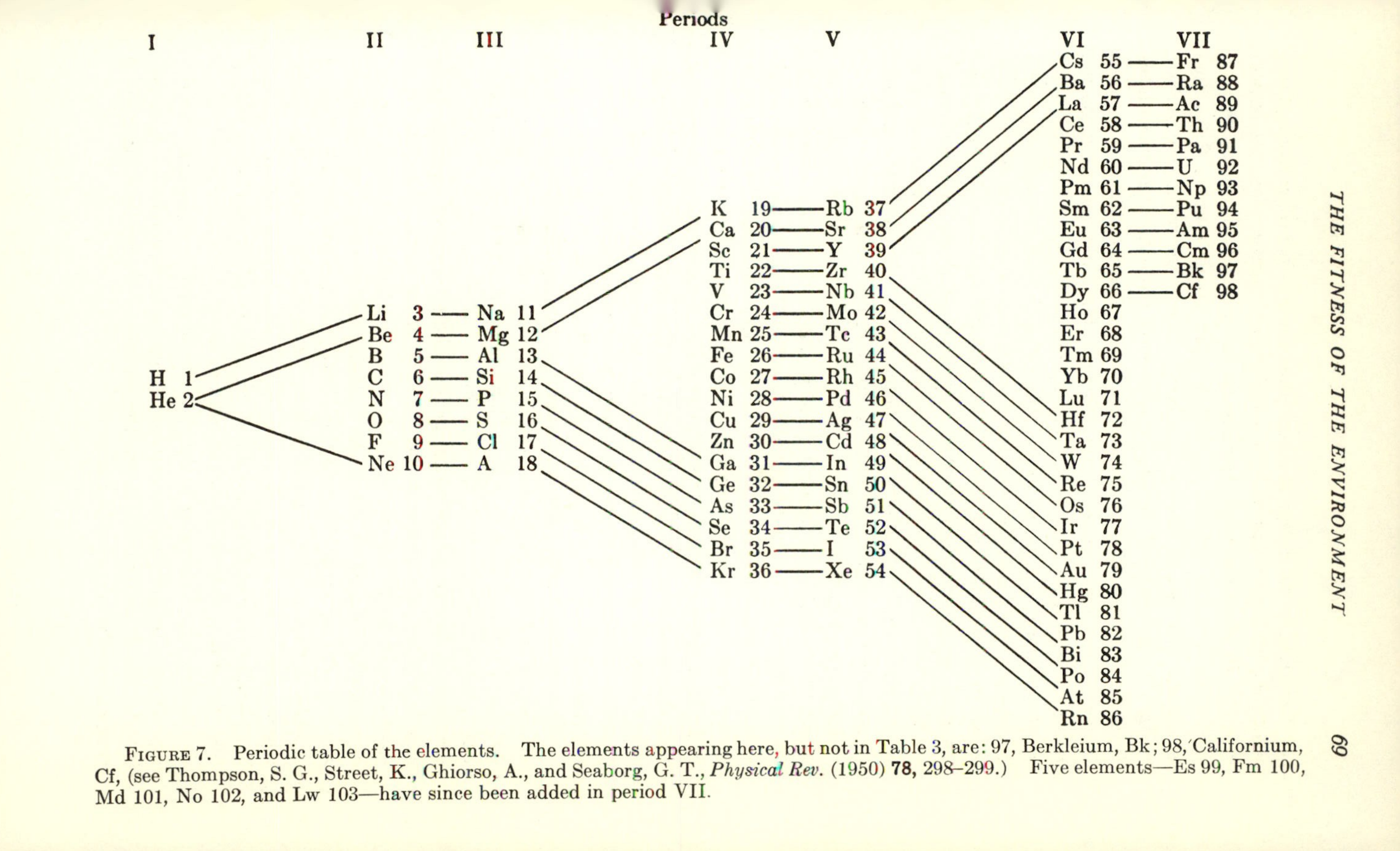

FIGURE 7. Periodic table of the elements. The elements appearing here, but not in Table 3, are: 97, Berkleium, Bk; 98, Californium, Cf, (see Thompson, S. G., Street, K., Ghiorso, A., and Seaborg, G. T., *Physical Rev.* (1950) **78**, 298–299.) Five elements—Es 99, Fm 100, Md 101, No 102, and Lw 103—have since been added in period VII.

Referring to the table, it is seen that the first period consists of two elements, hydrogen and helium, with atomic numbers 1 and 2 respectively. In these elements there is only one shell with places for only two electrons, and in helium, which with atomic number 2 has two electrons, this shell is filled. Similarly, each of the remaining periods ends in a noble element in which, as in helium, the number of electrons is just sufficient to fill the outer shell; the inertness of these elements is associated with a completed outer shell. The hydrogen atom, on the other hand, having only one electron, and hence an incomplete outer shell, can form molecules by combining with other atoms, a general property shared with all the other elements with incomplete shells. Next in order after helium is lithium with atomic number 3; two charges on the nucleus are taken care of in filling the first shell which we have seen to be complete in the helium atom, leaving one electron in the outer shell. Similarly, the other alkali metals all have only one electron in the outer shell, while in contrast the halogens all have only one electron less than is required to complete the outer shell. Only elements belonging to the first three periods will be discussed here, so the arrangement in the later periods, which becomes somewhat more complex, need not be dealt with. It is obvious that the chemical properties of an element are associated with the number of electrons in its outer shell, which in turn depends upon the nuclear charge represented by the atomic number. It is these electrons of the outer shell that are involved predominantly in forming the chemical bonds that hold atoms together in molecules, and they are often referred to as the valence electrons.

A molecule always has less total energy content than that of the atoms from which it is formed, the difference in energy representing the energy of the chemical bonds. There are different types of chemical bond, involving the valence electrons in different ways. There is a tendency to complete the outer shells of the atoms so as to form stable configurations analogous to those of the noble gases. In the *covalent bond*—which is the most important type, particularly among organic compounds—a pair of electrons is shared in common between two atoms. For example, in methane (CH_4), carbon may be represented as sharing the four electrons of its outer shell with four hydrogen atoms which simultaneously share their single electrons with the carbon. This results for carbon in a stable configuration comparable to that of the neon atom, with eight electrons in its outer shell, there being two in the inner shell. The four hydrogens at the same time assume the configuration of helium, each with two electrons in a single shell. Employing dots to represent electrons, the structure of methane, neon, and helium may be pictured as follows:

H
..
H: C: :H   : Ne: :   He:
..
H

methane   neon   helium

More commonly, only the electrons of the outer or valence shell are indicated,

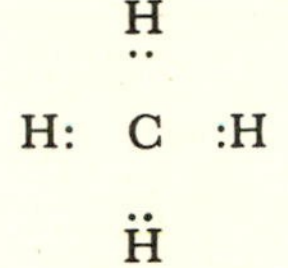

and often the covalent bond is represented simply by a line,

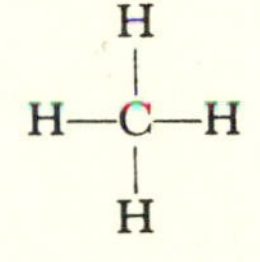

Another type of bond involves the forming of charged ions, or the tendency thereto. For example sodium, with 11 positive charges on the nucleus and 11 electrons, may loose one electron from its outer shell, and thus achieve the stable configuration of the noble element neon, with 10 electrons; it thus becomes a stable *ion* with a positive charge since there is one more positive charge on the nucleus than there are electrons in the shell. In the same way chlorine with 17 positive nuclear charges may gain an electron to become a stable ion with 18 electrons like the element argon, a negative charge being given it by its extra electron. In a solution these ions remain stable and independent, but when brought close to each other as in a crystal of common salt (NaCl) they form a bond due to the mutual attraction of their electrical charges. Such a bond is called an *ionic bond*. The tendency to form ionic bonds is measured by the relative electronegativity of the elements involved, which is illustrated in Figure 8, and is seen to be related to their positions in the periodic table. In general, the farther apart the elements are in the electronegativity scale, the more likely they are to form ionic rather than covalent bonds. Also, the greater the separation on this scale the stronger the bond, that is, the greater the energy which is lost by the molecule in forming the bond.

Purely ionic bonds are not common, but many chemical bonds are partly of this character, and partly covalent. The intermediate nature of many chemical bonds is now explained by the theory of resonance, which is supported by the reasoning of quantum mechanics. According to this theory, a molecule may enjoy two or more possible structures among which it resonates. Since the structures cannot be separated experimentally, the molecule and the bonds involved are best regarded as "hybrids" of two or more possible structures, each of which contributes a certain amount of the bond energy.

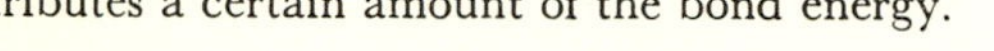

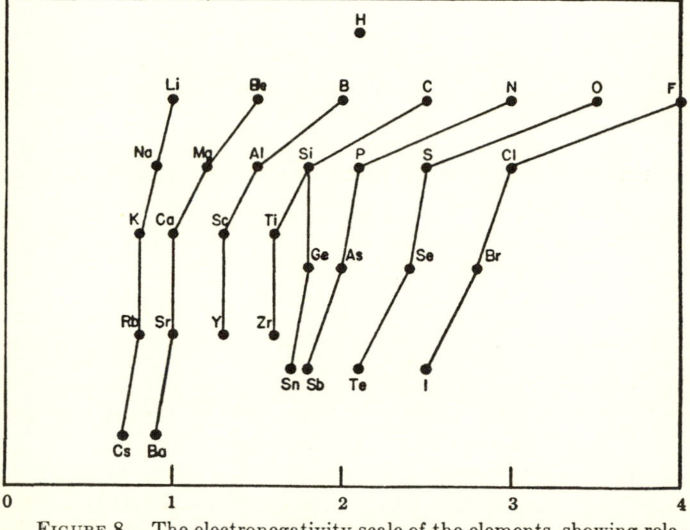

FIGURE 8. The electronegativity scale of the elements, showing relation to the periodic table. Abscissa, relative electronegativity. Ordinates, elements arranged according to periods. (From Linus Pauling, *The Nature of the Chemical Bond*, Ithaca (1944), Cornell Univ. Press, by courtesy of the author and the publisher.)

A few other types of bonds are recognized in addition to those mentioned, including the hydrogen bond which will be referred to again shortly. For the present it is only necessary to point out that while the periodic table gives a good general index of the properties that elements and their compounds may be expected to have, because these properties derive from atomic structure which is also the basis of the periodic table, one must go somewhat further to gain an understanding of the relations which determine fitness. No doubt there are still pertinent facts to explain, but in general Henderson's thesis may be more effectively supported in terms of modern chemistry than was possible when *The Fitness of the Environment* was written.

The Hydrogen Bond and the Fitness of Water

There is a type of bond that may be formed only between hydrogen and those few other elements which enjoy greater electronegativity (see Fig. 8). It is a bond of low energy—5 to 10 kilocalories per mole—in which the H atom behaves as if it were shared between two other atoms; hence it is sometimes called a hydrogen bridge, although it is more commonly known as the *hydrogen bond.* The exact nature of this bond is debated, but it appears to possess more of ionic than of covalent character. Our interest focusses in the present instance on the ability of this bond to associate or link together the molecules of water and of analogous compounds; the bond may be represented, without commitment as to its exact nature, for water, hydrogen fluoride and ammonia, respectively, as follows:

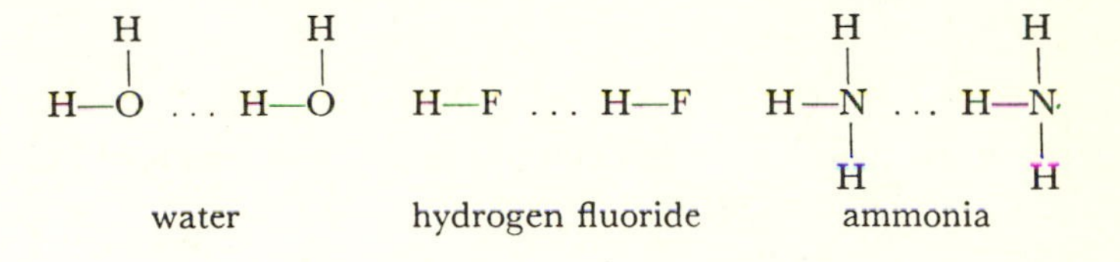

the dots representing hydrogen bonds between the molecules.

Figures 9 and 10 permit a comparison of three properties of various hydrides. The three properties, melting point, boiling point, and latent heat of vaporization have already been used in this book to illustrate the fitness of water. In each case the melting points of the hydrides of a given group of elements—e.g. for the halogens, HF, HCl, HBr, and HI—are plotted in the order of their atomic numbers and their places in the periodic table. It is seen that three substances, water (H_2O), hydrogen fluoride (HF), and ammonia (NH_3) do not fall into line with the corresponding hydrides of the other members of their series. Taking, for example, the melting points of the halogens, it is noted that for the hydrides HI, HBr, HCl, and HF this property decreases in regular order with decreasing atomic number until HF is reached. That substance instead of melting at about −150°C, as would be predicted by extrapolation from the other members of the series, has a much higher melting point, −92.3°C. Again in the case of the boiling point, for the series of which water is a member, we find that the boiling point falls regularly through H_2Te, H_2Se, H_2S, and extrapolation would place that of water at about −80°C instead of its true value, 100°C. In the case of H_2O, HF, and NH_3, and only in these three compounds, the properties examined are much higher than would be predicted; for we note that the hydride of carbon CH_4 falls more or less in line with the higher members of its series. These abnormal properties of the hydrides of nitrogen, oxygen, and fluorine are due to the ability of these elements to form hydrogen bonds, since extra energy is required to break these bonds in undergoing the change of state involved in melting, boiling, or vaporization. Figure 8 shows that these are the elements most likely to form such bonds, having the highest electronegativities relative to H.[5]

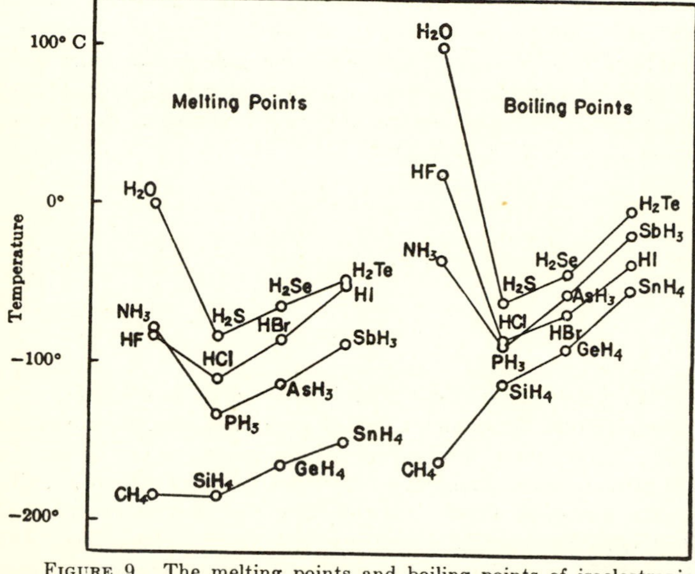

FIGURE 9. The melting points and boiling points of isoelectronic sequences of hydride molecules. Compounds arranged along the abscissa according to the groups to which the element combined with H belongs, e.g., O, S, Se, Te; and F, Cl, Br, I, etc. (From Linus Pauling, *The Nature of the Chemical Bond*, Ithaca (1944) Cornell Univ. Press, by courtesy of the author and the publisher.)

In general, boiling point rises with increase in atomic and molecular complexity, and melting point follows boiling point somewhat irregularly. Since atomic complexity increases with atomic weight, the heavier elements and their compounds tend to have higher melting points and boiling points. If we picture the nascent earth as cooling from a considerably higher temperature than its present, we may ask

[5] The existence of the hydrogen bond was only recognized after the publication of *The Fitness of the Environment*. Its importance was pointed out by W. M. Latimer and W. H. Rodebush in 1920 (*J. Am. Chem. Soc. 42*, 1419–1433), but was not generally recognized until considerably later. Knowledge of this property of hydrogen would have relieved Henderson's task considerably.

what effect this had on the evolution of the earth. The heavier elements and their compounds should have tended to liquify and then to solidify before their lighter companions, and as we have seen the former are in general those substances that make up the solid mass of the earth. No matter what ideas we choose as regards the evolution of the earth, it is obvious that any substance which was to remain liquid or gaseous at

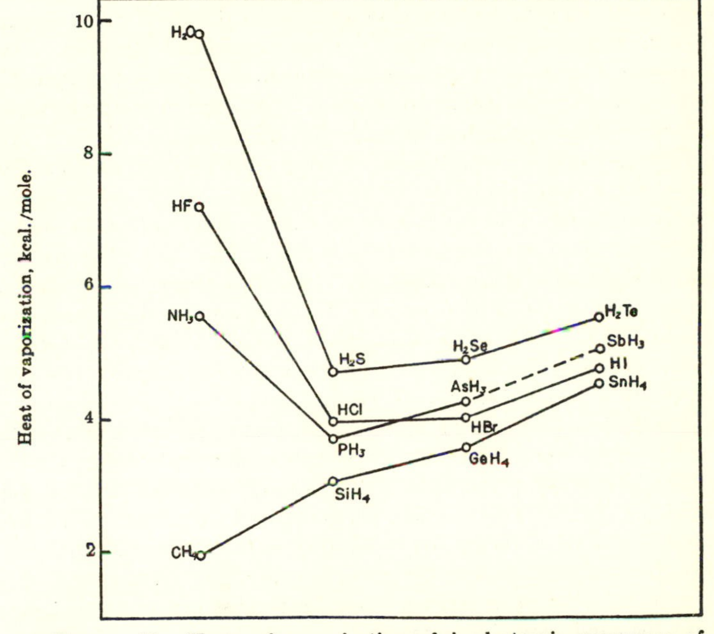

FIGURE 10. Heats of vaporization of isoelectronic sequences of hydride molecules. Arrangement of compounds the same as in Figure 9. (From Linus Pauling, *The Nature of the Chemical Bond*, Ithaca (1944) Cornell Univ. Press, by courtesy of the author and the publisher.)

the temperature of the earth's surface would need to be made up of the lighter elements.

Henderson recognized that ammonia shares many of the properties that contribute to the fitness of water, and suggested that the former compound might play a corresponding role on some planet with a much lower temperature than that of the earth. He did not apparently have available the data which would have shown that at some intermediate temperature hydrogen fluoride might have been quite as fit. Considering this point more carefully, we may come to a better appreciation of the meaning of fitness. Let us, for example, consider the property, high heat of vaporization, the importance of which in the case of water has been pointed out. This aspect of fitness is shared by all the three hydrides H_2O, HF, and NH_3, and a common cause has been demonstrated in the hydrogen bond. As far as this aspect alone is considered, any one of these three substances would be "fit" if it were possible to choose the prevailing temperature so that the substance would be in the liquid state. But the temperature of the earth is not determined by factors directly related to those that determine the high vapor pressure of water, but by such unrelated factors as the earth's rate of rotation and its distance from the sun. If we care to go so far back in the stream of causality, we may have to admit that it is those "accidental" factors which set the temperature of the earth, and hence make water more fit in this respect than its analogous hydrides.

Perhaps the argument appears at this point to become circular, but we can try to break out of the circle by imagining water, say, to possess properties slightly different than those it has. If we do, we are immediately brought back to the circle by the obvious fact that the sum total of properties of any substance, including water, is peculiar to that substance. So fitness partakes of the nature of uniqueness, and this uniqueness of the earth as an abode of life is a matter that strikes one more forcibly the more he tries to break out of the circle. Not only is the earth as it is, but it has reached that state through an evolutionary process, each step of which has been dependent upon the one preceding it. The stage upon which living systems bowed their debut was set by all the preceding events in the history of the earth—or, for that matter, of the universe. These events placed important restrictions upon the nature of life and its evolution. Life, it seems, did not arise and evolve as a system free to vary in any direction whatsoever; but as a system upon which great restrictions were placed, some of them even before the earth came into existence.

The anomalous behavior of water in the neighborhood of 4°C is also associated with hydrogen bonds; and the great solvent power of water and its ability to produce ions are associated with an abnormally high dielectric constant resulting largely from the formation of these bonds. Thus the hydrogen bond is seen to be a factor of great importance in determining the fitness of water as an environmental component. But the influence of this bond carries farther, to quote a chemist's opinion: "Because of its small bond energy and the small activation energy involved in its formation and rupture, the hydrogen bond is especially suited to play a part in reactions occurring at normal temperatures. It has been recognized that hydrogen bonds restrain protein molecules to their native configurations, and I believe that as the methods of structural chemistry are further applied to physiological problems it will be found that the significance of the hydrogen bond for physiology is greater than that of any single structural feature."[6] Some further suggestions as to the possible importance of the hydrogen bond in living systems will appear in later chapters, but we have already seen that this unique property of the element hydrogen plays a dominant role in determining important characteristics of the environment in which living systems have evolved and in which they maintain themselves.

The Uniqueness of Hydrogen

All properties which contribute to the fitness of water are not directly attributable to the hydrogen bond, however. Its high heat capacity, for example, needs other explanation. To a certain degree of approximation, all solid elements have the same heat capacity per gram atom (law of Dulong and Petit); and the molar heat capacity of a solid compound is approximately equal to the sum of the atomic heat capacities of its constituents (Kopp's Rule). Henderson pointed out that if these rules hold, the elements with the lowest atomic weight should have the highest heat capacity per gram, and the same should apply to compounds which have large numbers of light atoms as constituents. Thus the compounds of hydrogen, including water, would be favored in this respect. But we are concerned here particularly with water in the liquid state, and the rules applying to gases and liquids are less exact than those for solids. Nevertheless, the suggestion has some justification in this case, for liquid water has an especially high heat capacity, and this, as has been seen, contributes to its fitness.

At this point, lest our ability to explain the fitness of water seem too perfect, the skeptic may be permitted to remind us that the transparency of water vapor at wavelength 10 μ, which has such great importance in regulating the temperature of the earth through the greenhouse effect, has not yet been explained in general terms. There may be numerous other examples of the kind that might be cited, but they could hardly compel us to abandon the idea of fitness.

We have seen that many of the properties that contribute to the fitness of water are really attributable to properties of hydrogen. Reference to the periodic table indicates the uniqueness of this element, which has little resemblance to any other. Its atoms are the lightest and also the smallest, and the latter property permits it to fit into configurations requiring a volume too small for other atoms. This and its central position in the electronegativity scale contribute to the large number of chemical combinations into which hydrogen may enter—

[6] Pauling, L. *The Nature of the Chemical Bond*, 2nd ed., Ithaca, N. Y. (1944) Cornell Univ. Press.

the greatest for any element. Yet the existence of this element on the earth is, or once was, rather precarious because its low atomic weight permits its loss from the earth's gravitational field. Probably great amounts of hydrogen have been lost, and that which remains was kept only because of its ability to combine with other elements to form molecules of sufficient mass to be held. Had our planet been a little smaller, or a little hotter because nearer to the sun, there might not have been enough hydrogen remaining to play its role in the evolution of living organisms— a suggestive thought to weigh in considering the uniqueness of the earth as an abode for life.

The Fitness of Carbon

The element carbon is second only to hydrogen in the number of its known compounds, oxygen being third, and these three elements are also by far the most plentiful in living systems. Table 4, showing the number of hydrides of carbon, illustrates the variety of combinations of this element. The central position of carbon in the periodic table—it can achieve a stable configuration by either losing or gaining four electrons, and characteristically forms four bonds with other elements—and its ability to form strong carbon-to-carbon bonds and hence long chains and rings, contribute to its large number of chemical combinations. The carbon-to-carbon bonds make possible a great variety of arrangements of the atoms in molecules, with corresponding variety of chemical properties, yet withall a degree of order, since similarly arranged molecules are generally similar in their properties.

Table 4.[1]

Element of Period II	Li	Be	B	C	N	O	F	Ne
Number of hydrides	1	1	7	~2300	6 or 7	2	1	0

[1] After Williams, R., *Introduction to Organic Chemistry*, New York (1927), D. Van Nostrand.

As simple examples; the two following structures have certain properties in common, because they both possess the OH structure. Both are alcohols:

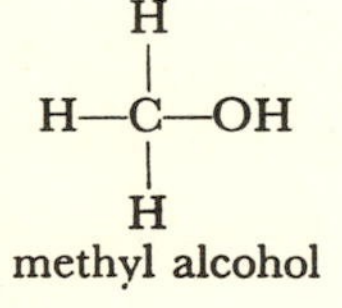

methyl alcohol

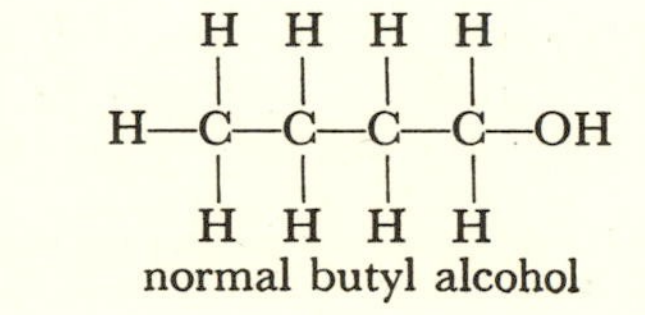

normal butyl alcohol

Another arrangement of the atoms in butyl alcohol results in somewhat different though similar properties:

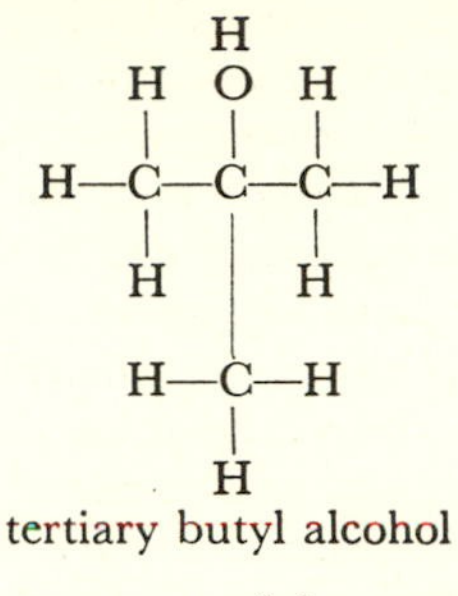

tertiary butyl alcohol

A somewhat different arrangement of the same atoms, results in a compound with quite different properties, common ether:

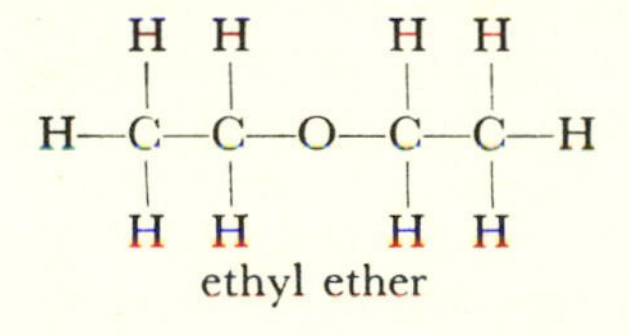

ethyl ether

As much larger molecules are formed, such as those characteristic of living systems, the variety of possible arrangements within the molecule increases enormously, and the possibility appears of having nearly identical molecules, which can differ in numerous minor ways while preserving the same general properties. This allows wide variations on a general theme of carbon compounds, which characterize living organisms in general and provide the working basis for evolution by variation and natural selection—but this is a subject for later chapters.

A comparable position in the periodic table is enjoyed by silicon, the congener of carbon in the third period (see Fig. 7), and it is sometimes asked why this element is not more generally found as a component of living organisms, playing an important role similar to that of carbon. Silicon can be made to enter into a large number of complex compounds in which it characteristically forms silicon to oxygen bonds; in fact this element is largely responsible for the great variety of minerals found in the earth's crust, and perhaps if we existed at temperatures around 3,000°C, where SiO_2 is a gas corresponding to CO_2 at normal temperatures, we would be aware of a somewhat greater variety of compounds of this element.[7] But such speculation has little merit as regards life on this earth, since gaseous compounds of silicon cannot exist here, and, as will be seen shortly, the fact that CO_2 is a gas

[7] Whistler's remark quoted at the beginning of this chapter was allegedly apropos of his failure in chemistry at West Point.

at normal temperatures is a factor of great importance to the living organism, and contributes to the fitness of that compound.

It was the relationship of carbon dioxide to the living organism that originally led Henderson to a reconsideration of the question of fitness, for he recognized that while the organism is in various ways obligatorily adapted to this compound, natural selection could have had nothing to do with the universal occurrence of CO_2 in the organism. He wrote, "It is not possible to explain the significance of carbonic acid in this physiological process as chiefly adaptation; for natural selection can have nothing to do with the occurrence of carbonic acid in the living organism, or, presumably, with the nature of the original things upon the earth. [It was this obvious fact which originally led me to a reconsideration of fitness.][8] The presence of carbon dioxide is inevitable, and whatever the first forms of terrestrial life may have been, certain it is that carbonic acid was one of the constituent substances. From that day to this it has steadily fulfilled the function of regulating the reaction of protoplasm, and of body tissues and fluids."

Water and carbon dioxide provide the three principal elements for living systems, and through photosynthesis these compounds are combined to provide the principal reservoir of energy for such systems. The high solubility of the gas CO_2 in the liquid H_2O is an important factor in bringing the two substances together for the photosynthetic process. At ordinary environmental temperatures one volume of water in equilibrium with pure CO_2 at atmospheric pressure contains approximately an equal volume of this gas (0.9 volumes at 20°C) a much higher solubility than is displayed by most gases; whereas only 0.03 volumes of O_2 are dissolved under corresponding conditions. Ordinary air, however, contains much more oxygen than carbon dioxide, and when this is taken into account the quantities of CO_2 and O_2 in solution in water in equilibrium with air are found to be more nearly equal—about 4 and 6 cubic centimeters per liter, respectively. Regarded in an overall sense, CO_2 and O_2 are required in approximately equal quantities by plants and animals. These substances must reach the living cells through an aqueous medium, whether this be sea water or body fluid, and the approximately equal quantities of these two substances in water, when in equilibrium with air, seems to achieve some degree of fitness.

Dissolved CO_2 combines with water to form carbonic acid, H_2CO_3,

$$CO_2 + H_2O = H_2CO_3; \quad \Delta F = 2.01 \text{ kg-cal/mol} \qquad \text{(VI-1)}$$

From the small value of ΔF it is seen that this is a readily reversible reaction. The H_2CO_3 helps to determine the acidity of the medium,

[8] Henderson's footnote.

ionizing rather weakly to yield hydrogen ion H^+, and bicarbonate ion HCO_3^-,

$$H_2CO_3 = H^+ + HCO_3^-; \quad \Delta F = 8.81 \text{ kg-cal/mol} \qquad \text{(VI-2)}$$

When base is present, as is generally the case, the reaction

$$BOH + H_2CO_3 = H_2O + B^+ + HCO_3^- \qquad \text{(VI-3)}$$

may be assumed to take place, in which B represents whatever basic compound or radical is present; so that there is always a considerable amount of HCO_3^- available. Thus a balance results between H^+, HCO_3^-, and H_2CO_3, as represented in reaction VI-2, which may be expressed in the form,

$$\frac{[H^+][HCO_3^-]}{[H_2CO_3]} = K \qquad \text{(VI-4)}$$

where K is the equilibrium constant—called, in this case, the dissociation constant. Now from VI-1 we see that H_2CO_3 is proportional to CO_2, and so long as CO_2 is present it will always be drawn upon to provide H_2CO_3, so we may write,

$$\frac{[H^+][HCO_3^-]}{[CO_2]} = K' \qquad \text{(VI-5)}$$

where K' is the apparent dissociation constant. Rearranging in logarithmic form,

$$-\log [H^+] = -\log K' - \log \frac{[CO_2]}{[HCO_3^-]} \qquad \text{(VI-6)}$$

Now $-\log [H^+]$ is commonly used as an index of hydrogen ion concentration, being called pH, and $-\log K'$ is commonly known as pK, so we may substitute to obtain,

$$\text{pH} = \text{pK} - \log \frac{[CO_2]}{[HCO_3^-]} \qquad \text{(VI-7)}$$

which is the common form of the Henderson-Hasselbalch equation. pK may be determined experimentally, having the value 6.5. Hence from VI-7 we see that when the concentrations of carbon dioxide and bicarbonate ion are known, the hydrogen ion concentration, usually expressed in terms of pH, can be calculated; CO_2 may be determined as the amount of this gas which can be removed from the solution by means of a vacuum, and HCO_3^- as equivalent to the additional amount that can be removed after acidification. The Henderson-Hasselbalch equation is useful in studying respiratory exchange between air and body fluids of animals, but here we are concerned only with the apparent fitness of such an arrangement for the environment of living organisms. From VI-7 it is seen that when the concentration of HCO_3^- and CO_2 are equal, pH = pK = 6.5. Now since the pH of pure water, commonly used to define neutrality, is 7.0, the ratio of HCO_3^- and CO_2 can obviously vary quite widely without moving the pH very far from that of water—we say that the solution is *buffered* near neutrality. Living organisms are quite sensitive to hydrogen ion concentration, and most of them are killed if the pH moves very far from neutrality. The fitness of this buffering system seems obvious.

There are a good many weak acids that would buffer near neutrality, and in fact phosphates and proteins are quite effective in accomplishing this in some living organisms; body fluids usually have a pH nearer neutrality than 6.5, due to these buffer systems. But the CO_2-HCO_3^- buffer system has a particular advantage in that CO_2 is a gas; thus the system has a volatile component, which moreover is a "by-product" of the organism's metabolism, and is hence constantly renewable. The result is that if acid is introduced, say, as a metabolic product, the bicarbonate acts as a buffer to keep the hydrogen ion concentration near neutrality, and at the same time CO_2 tends to form more HCO_3^- so long as base is available. If excess base is introduced, on the other hand, the H_2CO_3 (supplied by CO_2) tends to neutralize it —again buffering near neutrality.

The Fitness of Oxygen

The fitness of the element oxygen has not been discussed thus far, although it is a member of the two compounds that have been chiefly used to illustrate fitness in general, H_2O and CO_2, and must contribute to the properties which determine this fitness. The name of this element is immediately associated with biological oxidation, the process whereby energy is made available to the living organism. I shall point out in the next chapter that while in an overall sense this process involves the uptake of O_2, in an intimate sense it more often involves the removal of H, with H_2O playing an intermediate rôle. Thus, again the fitness of oxygen is intimately related to the fitness of hydrogen and water.

As carbon and silicon were compared, so oxygen may be compared with sulfur, its nearest group neighbor in the periodic table. Reference to Figure 8 shows that oxygen should form hydrogen bonds more readily than sulfur, and the greater apparent fitness of H_2O as compared to H_2S may be largely accounted for on this basis. This factor alone might be expected to establish greater fitness for oxygen than for sulfur. But oxygen has roles in living processes for which it may be more difficult to find rigidly specific fitness. O_2 plays a dominant part in the reactions of energy metabolism, yet it can be dispensed with by many anaerobic organisms. As a component of water, oxygen is concerned in photosynthesis, yet it is conceivable that sulfur might have taken its place. From an overall thermodynamic point of view H_2S could replace H_2O; indeed some photosynthetic bacteria do use the former compound, and selenium, another congener of oxygen and sulfur, may also take part in such processes. In known photosynthetic reactions, however, H_2S does not actually take the place of H_2O in the intimate mechanism, but the theoretical possibility remains, if one were reconstructing living organisms on another basis from that upon which they have actually originated and evolved.

The Fitness of Other Elements

The relative proportions of the various elements making up the human body are indicated in Table 5, which may be taken as representing, roughly, the composition of living organisms in general. A few additional elements—Rb, Cs, Li, Ba, Sr, Ag, and Cr, at least—are found scattered among the various plants and animals. The lack of direct relationship between amount and position in the periodic table, except for a general predominance of the lighter elements, is indicated in the table. Only in a very rough way does such a table suggest the importance of the various elements, as in the case of the first three items, hydrogen, carbon, and oxygen. The high proportion of the first two elements reflects the preponderance of water.

Compounds containing some of the elements other than hydrogen, carbon, and oxygen are known to play indispensable roles in living organisms, but the fitness of the particular elements is difficult to explain. For example, phosphorus is an essential component of substances concerned in the utilization of energy by the cell, and in this instance it is not known to be replaceable by any other element; but the reason for this unique physiological role is not so easily explained in strictly physicochemical terms. Similarly nitrogen and sulfur are universal components of proteins, and hence of all living systems, but why these elements have been elected for their particular roles is obscure. Numerous other elements take part in specific processes, but are not so universally required. For example magnesium is a component of chlorophyll, the most important pigment in photosynthesis, and iron is a component of hemoglobin and of numerous respiratory enzymes. Copper plays a role at least superficially analogous to that of iron in gaseous exchange and respiration in some animals. Numerous other elements, for example iodine, fluorine, and calcium, are specifically required in certain processes. The roles of sodium, potas-

Table 5. Elementary Composition of the Human Body (after Florkin, M., *Introduction a la biochimie générale*, 4th ed., Paris (1946), Masson et Cie.).

Element	Period (in periodic table)	Per cent of total weight	Per cent of total atoms		
H	I	10.2	63.5	98.2%	99.96%
O	II	66.0	25.6		
C	II	17.5	9.1		
N	II	2.4	1.06		
Ca	IV	1.6	0.25		
P	III	0.9	0.18		
Na	III	0.3	0.07		
K	IV	0.4	0.06		
Cl	III	0.3	0.05		
S	III	0.2	0.04		
Mg	III	0.05	0.01		
Fe	IV	0.005	0.006		< 0.1%
Zn	IV	0.002	0.0002		
Cu	IV	0.0004	0.00006		
Mn	IV	0.00005	0.00006		
Ni	IV	—	—		
Co	IV	—	—		
Al	III	—	—		
Ti	IV	—	—		
B	II	—	—		
I	V	—	—		
As	IV	—	—		
Pb	VI	—	—		
Sn	V	—	—		
Mo	V	—	—		
Vd	IV	—	—		
Si	III	—	—		
Br	IV	—	—		
Fl	II	—	—		

sium, and chlorine are somewhat less specific. Numerous elements are required in traces (e.g. cobalt and boron) in processes of more or less limited scope; that is, they are essential only within a limited part of the animal or plant kingdom. Fitness, in such cases, becomes more difficult to understand.

Fitness and Uniqueness

By way of summary of these ideas, we may ask to what extent fitness of the environment represents uniqueness of properties. Do the substances which appear so eminently fit owe their fitness to properties belonging only to those substances, or could they be replaced by other substances under appropriate conditions? We may find that the answer to our question is not always the same, and that the answer may depend upon the way in which the question itself is formulated. For illustration, let us recapitulate some examples of aspects of fitness:

1. Certain properties of water that contribute to its fitness are intimately linked with the properties of the element hydrogen. These properties depend upon the structure of the hydrogen atom, which is unique to that element and presumably the same no matter where the element finds itself in the universe—whether, for example, in the sun's photosphere or in the earth's hydrosphere—although of course the measurable properties may vary according to the surroundings. In this respect, fitness and uniqueness appear to be more or less synonomous.

2. On the other hand, when the substance water is considered, it is found that the same properties of hydrogen convey, in a parallel fashion, the same kind of properties to the substances HF and NH_3 which are found to give fitness to H_2O. An important factor enters now, however, which makes H_2O the only one of these substances that is actually fit as regards life on the earth; namely, the temperature of that planet, which does not permit HF or NH_3 to exist in the liquid state. It is not at all clear how water came to be present on the earth; obviously events were involved that did not occur, or occurred to only a limited extent, on our two nearest neighbor planets. But water's fitness as an environmental component may reside to a great extent in the fact that it was here in available form, and hence this aspect of fitness partakes of an historical or evolutionary quality, and is localized to this particular planet. This aspect of fitness is not, then, universal, but exists only in relation to the planet earth, or to planets that are very nearly like the earth.

3. The fitness of some elements that play apparently unique roles in living systems, e.g. nitrogen, phosphorus, and sulfur, seems to depend still less upon properties unique to the particular element.

This is no doubt due in part to the complexity of the processes in which they participate, and our inability to grasp the immediate relationships. On the other hand, it seems probable that in many cases the particular aspect of fitness is involved with the origin and evolution of living organisms. In such cases our understanding of fitness may be limited by our ignorance of essential steps in the history of life.

From these few examples it is seen that fitness and uniqueness are not strictly synonomous. Fitness is not to be defined in a few words, and perhaps only rather vaguely at best, being, as it is, a concept involving a complex of factors. Obviously one must be cautious in pushing the concept too far, lest like Doctor Pangloss he picture "le meilleur des mondes possibles," in which all parts are perfectly fitted to all others. The value of the concept is quite the reverse. Taken in moderate doses it leads to the opposite point of view, serving to illustrate the existence of limitations to the range of properties of living systems, and to show that the direction of organic evolution must have been guided to a considerable extent accordingly.

VII · THE ENERGETICS AND KINETICS OF LIVING SYSTEMS

"*Die Pflanzen nehmen eine Kraft, das Licht, auf, und bringen eine Kraft hervor. . . .*"—ROBERT MAYER

LIKE any other machine, the living system must have a supply of energy for its operation. If it does external work as, for example, in bodily movement or in the expulsion of waste products, free energy must be expended. Even more fundamental is the need for energy for growth and maintenance. How the contemporary organism obtains and utilizes energy is a fascinating problem of great importance; how the earliest living systems managed their energy requirements is an even more basic one from an evolutionary point of view.

A variety of chemical changes goes on continuously in the living organism, and to this the general term *metabolism* is applied. The mechanisms of accumulation and expenditure of free energy by living organisms are inseparable from this metabolism. Intensive study within recent years has added greatly to knowledge of the latter processes, but our understanding of the overall energy balance-sheet for living material as a whole remains essentially unaffected by these advances. A survey of this balance-sheet and its implications will be in order before passing to the consideration of more specific problems

The Overall Balance-sheet

Probably all living organisms, both plants and animals, contain in the normal state a certain reserve of "fuel" in the form of carbon compounds with high chemical potential from which free energy may be released. This supply being drawn upon for the various needs of the organism, the "energy-rich"[1] carbon compounds must be replenished from time to time by the capture of energy from the outside. In the long run, virtually all this energy comes from sunlight, through the intermediacy of photosynthetic plants.

[1] This adjective will be used to describe compounds or molecular configurations which may, in the course of a spontaneous chemical reaction, release a considerable quantity of free energy.

All organisms are energy "spenders" in the sense that they depend upon the utilization of self-contained supplies of energy-rich compounds, but only certain ones may be classed as energy gatherers. Those organisms that must be supplied with energy-rich carbon compounds are grouped under the term *heterotrophic.* They include virtually all the animals; and, among the plants, the fungi, myxomycetes, heterotrophic bacteria, and a few others. Those organisms which manufacture their own supply of energy-rich carbon compounds with no source of carbon other than CO_2 are referred to as *autotrophic;* and of these the algae and higher plants which carry on the process known as "green plant" *photosynthesis* contribute almost all of the energy-gathering side of the balance-sheet. The latter constitute much the greater part of plant life, in mass if not in number of organisms. For purposes of an overall energy balance-sheet, then, one may separate two groups, the energy spenders, composed of the heterotrophs, and those which gather the energy of sunlight, the photosynthetic plants.[2] The latter not only supply their own needs for energy-rich compounds, but since it is their fate sooner or later to be eaten by the former, they furnish the free energy for all living processes.

Energy capture by green plant photosynthesis may be represented by the schematic reaction

$$\text{carbon-dioxide} + \text{water} + \text{sunlight} \rightarrow \text{energy-rich carbon compounds} + \text{oxygen} + \text{heat} \quad \text{(VII-1)}$$

Correspondingly, the process of expenditure of energy may be schematized as follows,

$$\text{energy-rich carbon compounds} + \text{oxygen} \rightarrow \text{carbon-dioxide} + \text{water} + \text{free energy} + \text{heat} \quad \text{(VII-2)}$$

This type of reaction may be taken to represent *biological oxidation*, the oxidation of energy rich carbon compounds with the release of free energy; chemically it is the reverse of VII-1. To include all types of biological oxidation the scheme would require numerous additions and

[2] The fact that some heterotrophic organisms can manufacture energy-rich carbon compounds does not class them as energy gatherers in our balance sheet, since they must utilize energy-rich compounds to do so. The chemoautotrophic or chemosynthetic organisms utilize energy from inorganic reactions, but probably contribute only slightly to energy gathering, and besides at least a part of the inorganic compounds they use may result from biological activity. The photoautotrophic or photosynthetic bacteria, while true energy gatherers, likewise contribute only a small amount to the balance-sheet. The various autotrophic forms will be of considerable interest in the discussion at the end of this chapter but they may be regarded as negligible quantities in the immediate discussion of the overall balance-sheet.

alterations, but it will serve for the moment to describe the overall energy exchange. Both of these processes are compatible with thermodynamic reasoning. In the first, radiant energy is used in the manufacture of energy-rich compounds, but since a part of the energy taken in is lost as heat the process is less than one hundred per cent efficient. In the second process the energy-rich compounds formed in the first are made to yield their energy—a part as free energy which may be made to accomplish work, and a part as heat. Again the process is less than one hundred per cent efficient, so this is not a perpetual

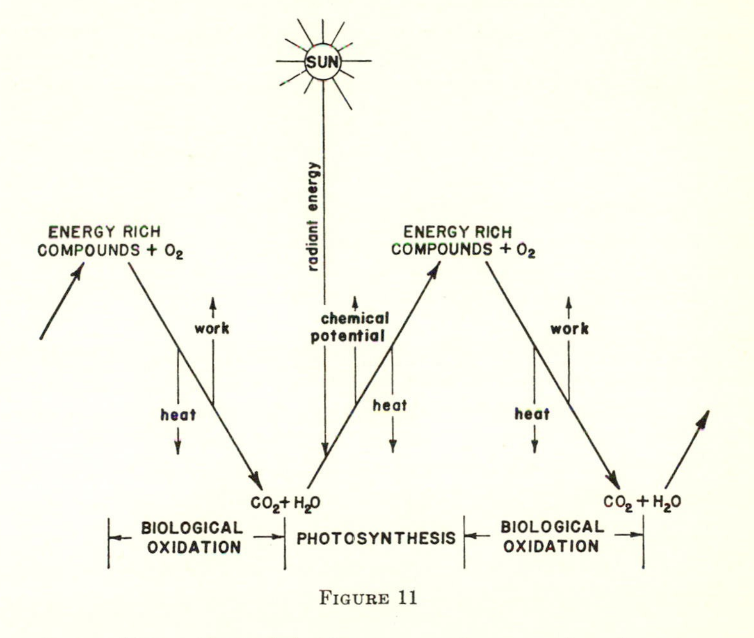

FIGURE 11

motion machine. Regarded either separately or together, the schemes are thermodynamically sound.

The overall process may be regarded as a cycle, as diagrammed in Figure 11. In the oxidative phase, which goes on in both plants and animals, energy-rich carbon compounds are represented as oxidized to CO_2 and H_2O, the reaction going "downhill" on the free energy scale. An arrow pointing upward indicates that in this phase of the cycle work is done (this includes growth and maintenance), and another arrow pointing downward indicates that heat is produced. In the photosynthetic phase, radiant energy from sunlight supplies the energy for building energy-rich compounds, thus increasing chemical potential, as indicated by the upward pointing arrow; but, again, a downward pointing arrow indicates that a part of the energy captured in the photosynthetic process goes to heat. In a complete balance-sheet, both phases of the cycle are included, sunlight being the ultimate source of the energy which drives the whole machine.

Analogy to a hydroelectric plant, diagrammed in Figure 12 may help towards understanding this balance-sheet. Water falling from a reservoir is made to drive a turbine which in turn operates an electric dynamo. The energy for driving the dynamo is derived from the energy lost by the water in falling from a higher to a lower gravitational potential, some fraction of which is converted into electrical energy by means of the turbine and dynamo. A part of the electrical energy so

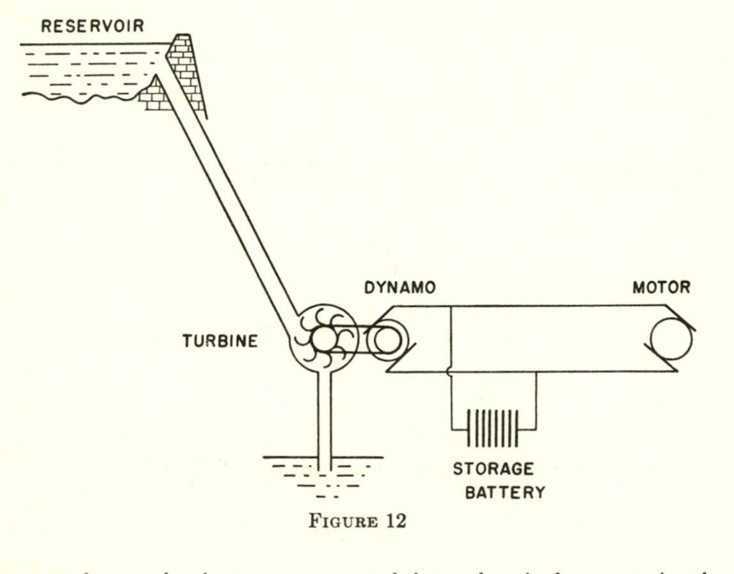

FIGURE 12

generated may be in turn converted into chemical energy in the storage battery. The photosynthetic phase of the cycle diagrammed in Figure 11, may be regarded as analogous to this much of the hydroelectric system, the energy for photosynthesis being derived from radiant energy lost by the sun. The oxidative phase is analogous to the operation of a motor by the storage battery, the motor accomplishing work. If the hydroelectric plant were cut off, the motor could continue to run so long as the energy stored in the battery lasted, this representing the situation of the heterotroph which possesses a limited supply of stored energy, but is ultimately dependent upon the plant. It also represents the green plant during its nonphotosynthetic periods.

The two processes represented above by schematic equations VII-1 and VII-2 might be described by a more formal chemical equation which is also schematic,

$$C_6H_{12}O_6 + 6O_2 = 6CO_2 + 6H_2O$$
$$\Delta F = -688 \text{ kg-cal/mol}$$
$$\Delta H = -673 \text{ kg-cal/mol} \qquad \text{(VII-3)}$$

Reading from left to right in the usual manner, this reaction describes the combustion of glucose,[3] and may be taken to represent biological oxidation; reading from right to left it would describe the synthesis of glucose from carbon-dioxide and water, and may be taken to represent photosynthesis. Both reactions occur in living organisms, but go through a series of steps which are not indicated here, and from the standpoint of mechanism some of the intermediate steps may be more typical of the general process than are the overall reactions. Nevertheless, the scheme is useful for illustrating the energy balance-sheet, which cannot be drawn exactly in any case.

Reading from left to right (biological oxidation) ΔF has a very large negative value, and hence the reaction can go spontaneously, and virtually to completion. The maximum amount of useful work that can be got out of the reaction is 688 kg-cal/mol, the value of ΔF; but in the course of biological processes a considerable part of this goes to heat. Reading from right to left (photosynthesis) the sign of ΔF is positive, showing that energy must be put in if the reaction is tc go in this direction. That the energy exchange can be represented as reversible in the above scheme does not mean that the two phases, photosynthetic and oxidative, must exactly balance. In fact, in the long run photosynthesis probably exceeds oxidation, since the latter cannot continue without the former for a very long period; and some energy derived from photosynthesis is stored in coal and petroleum. There is probably a fairly close balance between the two processes at the present time.

Any prolonged imbalance between photosynthesis and oxidation should be reflected in the ratio of the two atmospheric components CO_2 and O_2, since according to reaction VII-3 the amount of CO_2 used up in photosynthesis is equal to that freed in a corresponding amount of oxidation. Thus, if photosynthesis exceeded oxidation, O_2 should increase in the atmosphere and CO_2 decrease.[4] At present our atmosphere contains, by volume, about 21 per cent of O_2 and only 0.03

[3] Glucose is one of a number of hexose sugars which are isomers having the same empirical formula.

[4] Coal and petroleum contain lower proportions of oxygen with respect to carbon than is indicated in equation VII-3, so their formation or combination should have a somewhat greater effect on the CO_2 content of the atmosphere than on the O_2 content.

per cent of CO_2; the total amounts are about 2.1×10^{19} moles and 4.9×10^{16} moles, respectively.[5] Estimates for total photosynthesis correspond to the reduction of about 1.5×10^{16} moles of CO_2 per year,[6] at which rate it would take only a few years to exhaust the supply of atmospheric CO_2 if none were returned by oxidation. Even with this rapid rate of turnover, however, small imbalances between photosynthesis and oxidation would not be detectable in the atmosphere as a whole, since oxidation follows photosynthesis fairly closely over any extended period. Local decrease in CO_2 may be readily demonstrated when active photosynthesis goes on in densely vegetated areas in the daytime, but the balance tends to be rapidly adjusted during darkness. Over the great periods of time concerned in earth history, on the other hand, the constitution of the atmosphere could have been considerably modified by an excess of photosynthesis over oxidation. Such alteration of the atmosphere has been assumed in various hypotheses regarding the evolution of the earth's surface and of living forms. Indeed, it has been widely accepted that the atmosphere was virtually devoid of O_2 at one time, and that its release from CO_2 by photosynthesis is responsible for its presence there today. Recent reconsideration of this question suggests, however, that O_2 may have been present in quantity before the advent of living organisms, and that the present atmospheric oxygen had a source other than photosynthesis. At any rate, there must have been an adequate supply of O_2 for a long while before the beginning of an adequate paleontological record, since most of the organisms whose fossils appear at that time belonged to phyla that are today exclusively dependent upon atmospheric oxygen. This does not necessarily mean that the proportion in the atmosphere was as high then as it is today, since much biological oxidation can go on at considerably lower partial pressures of oxygen than is present in the modern atmosphere. The idea is also plausible that the atmosphere contained a high enough proportion of CO_2 to maintain the temperature at the earth's surface somewhat above what it is today, due to the "greenhouse effect," until the quantity of this gas was diminished by the excess of photosynthesis indicated by great deposits of coal during the late paleozoic era. Extensive deposits of calcareous material also indicate that much carbon has been fixed as carbonate at the surface of the earth, largely through the activity of living organisms. The temperature differences need have been only

[5] After Humphreys, W. J., *Physics of the Air*. New York (1940) McGraw-Hill.

[6] Based on estimates given by E. I. Rabinowitch, *Photosynthesis*, New York (1945) Interscience Publ. Some readers may be surprised to learn that the best estimates indicate that of the total photosynthesis for land and sea, the latter contributes several times as much as the former.

a matter of a few degrees to have profoundly affected the climate so far as living organisms were concerned.[7]

A considerable part of the sunlight is scattered by the atmosphere and hence never reaches the earth. Another large part falls on sparsely vegetated areas of land or water. A goodly fraction is reflected or transmitted by the plants themselves and hence does not participate in photosynthesis, and of the total absorbed radiation only a part can bring about photosynthesis. Thus the fact that only a very small fraction of the energy of sunlight is harvested in photosynthesis results from the small quantity utilized by the plants of the earth rather than low efficiency of the process itself. The net efficiency of photosynthesis may be quite high—up to sixty-five per cent is reported under experimental conditions. But the gross efficiency, which in our reckoning should include that energy required for building and maintenance of the photosynthetic machine—the green plant—must be very much lower.[8] It would be difficult, if not impossible, to estimate the overall efficiency of living systems from the absorption of sunlight to the accomplishment of mechanical work, but it could not be greater than a few per cent at most. The net efficiency for mechanical work of muscle may reach thirty per cent, but on the other hand the efficiency of excretion, as exemplified for example by the mammalian kidney, is not over a few per cent. The gross efficiency for these processes must be considerably lower, but is difficult to estimate in most cases. In fact, any estimate of gross efficiency must be in a sense arbitrary. How much of the energy used in growth and maintenance should be included? In the present instance, where we are concerned with the overall balance-sheet for all living systems, it would seem necessary to take into consideration all the energy used by organisms throughout their lifetimes, the integrated whole including the growth and maintenance of all

[7] The free energy increase involved in carbon dioxide reduction by photosynthesis amounts to about 1.8×10^{18} kilogram-calories per year, whereas the sunlight which falls upon the earth's profile during the same period is about 1.27×10^{21} kilogram-calories. Thus, only about 0.14 per cent of sunlight is made available for life processes, and any conceivable imbalance between photosynthesis and oxidation could have little direct effect on the temperature of the environment as a whole.

[8] The efficiency of any machine is defined by the relationship

$$\text{Efficiency} = \frac{\text{Output of Work}}{\text{Input of Energy}} = \frac{\text{Work}}{\text{Heat} + \text{Work}}$$

Net and gross efficiency are comparable to net and gross income, and their calculation is somewhat arbitrary. As an example, *gross* efficiency of a man riding a bicycle is about 20 per cent after correction for losses in friction of the machine. If correction is made for processes not directly concerned in muscular action, such as basal metabolism and friction of the muscles, the *net* efficiency for muscular work of bicycle riding is about 33 per cent (Benedict, F. G., and Cathcart, E. P., *Muscular Work*, Publ. Carnegie Inst. Wash. No. 187 (1913)).

living things over any given period of time.[9] No matter how we view the matter there is no evidence of controversion of the first law of thermodynamics; in terms of total energy exchange living systems are not very efficient.

The problem posed in considering the gross efficiency finds its parallel in applying the second law of thermodynamics. Living organisms represent systems that are highly "organized," that is, they display less randomness than the materials from which they are "built"; and it is therefore justifiable to say that a decrease in entropy is involved in their building. This seems to be the basis for occasional statements to the effect that living organisms are not governed by the second law of thermodynamics. But such a point of view could only be justified by showing that the decrease in entropy involved in building the organism is not accompanied by a greater increase in entropy of the surrounding universe. This seems most unlikely, and certainly not demonstrable, since in the course of building the organism a great deal of energy must flow through it—just how much may depend, as in the case of the gross efficiency, upon the point chosen to represent the beginning of the building process. The small local decrease in entropy represented in the building of the organism is coupled with a much larger increase in the entropy of the universe.

The temperature of the sun is near 6,000°K, that of the earth about 285°K. Energy flows from the former to the latter with an accompanying increase in entropy, and a suitable machine can convert into work some of the energy flowing from the hot to the cool body. Analogy to the hydroelectric model used above is obvious, and analogy may also be drawn to the steam engine, where the difference in temperature between the boiler and the exhaust corresponds to the difference in temperature between earth and sun. To come closer to the true case, the radiation of sunlight and the black-body radiation from the earth may be compared. The spectral distribution corresponds to the respective temperatures of the two emiting bodies, the maximum for the sun's photosphere lying at wavelength 0.48 μ, that for the earth at 10 μ. Radiant energy is composed of discrete quanta, the magnitude of the individual quantum being inversely proportional to the wavelength of the radiation.[10] For the maximum wavelength of sunlight the quantum is 4×10^{-12} ergs, for the maximum of the earth's black body radiation 0.2×10^{-12} ergs. The total radiant energy received by the earth from the sun is virtually equal to the energy

[9] This would be like including the energy expended in building and repair of an automobile in the calculation of the efficiency of its motor—an unusual engineering procedure, but one which seems necessary in order to gain a proper thermodynamic evaluation of living systems.

[10] Equation VII-21 and accompanying discussion.

radiated by the earth to space. But the quality of the two radiations is different, the former being composed of a smaller number of larger particles of energy. One might say, then, that the entropy of sunlight is less than that of the earth's radiation, because the former displays less randomness. Looked at in this way the overall process involves a great increase in entropy and hence the capture of a little free energy accompanied by a small local decrease in entropy by living systems is not in any way contrary to the second law of thermodynamics. The entropy and free energy of living systems may be considered to be approximately constant; there is a flow of energy through them, but very little is stored for long. No matter how one regards the overall energy relationships, he finds nothing to justify the concept that living organisms behave in a way contrary to the thermodynamic principles that apply to nonliving systems.[11]

The Expenditure of Free Energy

The small fraction of the energy of sunlight which is captured by living systems is stored for longer or shorter periods in carbon compounds as chemical potential. This chemical potential is then released as free energy in the course of exergonic chemical reactions, a certain portion being utilized for various activities of living organisms. The mechanism of capture of the energy of sunlight will be taken up in a later section, but it will be convenient to discuss first the expenditure and utilization of the stored energy. This is comparable to considering the thermodynamics of a steam engine without taking into account the source of the energy released in the combustion of the coal, which of course stems from the same ultimate source, sunlight.[12] In both cases the energy is released by chemical reactions which belong to the type known as oxidations, but the course of these reactions in *biological oxidation* is very different from the burning of fuel.

Under appropriate conditions one might utilize the combustion of glucose for heating a steam boiler to operate an engine. In that case, the energy obtained for heating the boiler would correspond approximately to the value of ΔH given in equation VII-3; of course only a

[11] Some years ago (*Science* (1937) *86*, 285), to emphasize the dependence of living systems upon sunlight, I described life as a "photochemical steady state." Today, speaking in the same vein, one might describe life as fed by a nuclear reaction, for in the meantime it has been learned that this is the source of the sun's energy. Some may find intriguing the idea that the atomic energy which man is now endeavoring to harness to his needs and whims has actually nourished him and all his ancestors since the remotest times—for at least a billion years.

[12] This is sound engineering practice, although we are beginning to see that it is not sound long-view economics. It is sound laboratory practice to study biological oxidation without regard to the ultimate source of the energy, but it does not give us a correct view of life processes as a whole.

certain fraction of this energy can be got out of the engine as work. In an overall sense, the same reaction describes the oxidation of glucose by a living organism. Allowing for differences in the temperature at which the reactions go on, the overall energy exchange for equivalent amounts of glucose is approximately the same in the two cases, if the conditions of the reactants and products are the same. But here the analogy ends, for the oxidation of glucose takes place in a very different way in the living organism, where it proceeds at a temperature at which pure glucose and O_2 could remain together indefinitely without reacting.

Although the reaction of glucose with O_2 to form CO_2 and H_2O has served as a useful scheme in considering the overall energy balance-sheet, it should not be thought of as describing accurately the reactions that go on in living systems. Many living organisms are indeed able to oxidize glucose, and this reaction has been the subject of a multitude of biochemical studies, but these show that the reaction is carried on in a series of steps. Glucose is a fairly complex compound, as the following structural formula indicates:

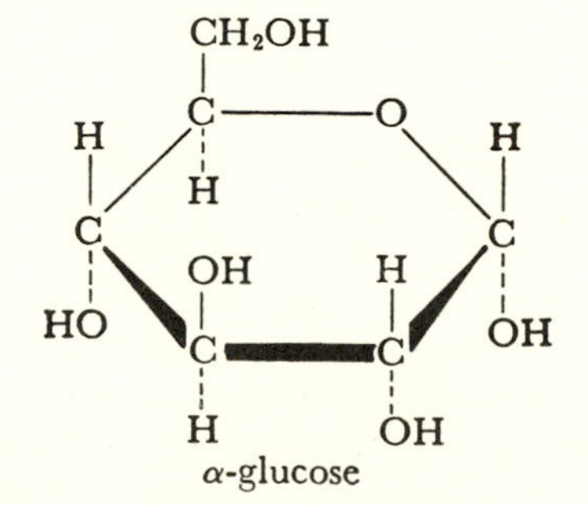

α-glucose

The six membered ring is to be thought of as lying in a horizontal plane, the solid vertical lines represent bonds above, the dotted lines below this plane. This substance may be broken down through a series of intermediate reactions involving less complex compounds until the end products CO_2 and H_2O are ultimately reached. Thermodynamically the energy changes in these intermediate reactions add up to the same total as if the reaction went as directly as indicated in VII-3. The intermediate steps themselves would go very slowly—some not at all—without the intervention of special catalysts, the *enzymes*, and other participating substances, the *coenzymes*. All of the former and many of the latter are complex substances found only in living systems. The obtaining of free energy by living organisms is associated with the free energy changes involved in some of these intermediate steps rather than with the overall oxidation of glucose.

It might be expected that living organisms would be able to employ the oxidation of some of these simpler intermediate compounds for obtaining their free energy requirements, and such is the case. Quite a large variety of carbon-compounds are utilizable by living organisms, using this term in an inclusive sense, although given types are restricted more or less as to the kinds of compounds they may use. Many microorganisms, particularly species of bacteria, can grow on media containing only a single carbon compound which provides the whole of the free energy requirements, although some inorganic substances must also be incorporated into the process. The carbon compounds may be as simple as methyl alcohol,

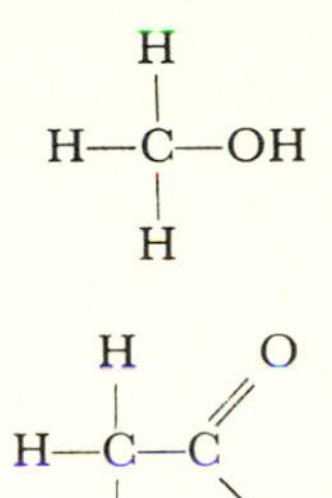

or acetic acid,

H O
H—C—C
H OH

All of the carbon compounds utilizable by living organisms for their free energy requirements are not directly derivable from glucose itself; a complete list would include, for example, a variety of nitrogen-containing compounds. Classically the higher vertebrate animals are supposed to use only three types of compound to satisfy their energy requirements—carbohydrates, fats, and proteins. These are broken down into simpler units, e.g. proteins into amino acids, fats into fatty acids and glycerol, before oxidation proper begins. For a long time, interest focussed on this type of metabolism, which includes that of man, but at present the metabolism of bacteria and other of the lower forms is receiving much attention, and even for the vertebrates the picture is shifting. As a result the study of metabolism gains wider scope, revealing a multiplicity of types through which runs a common pattern, suggesting evolutionary relationships. Here no attempt will be made, however, to review this rapidly advancing field.

Organisms do not always carry out the oxidation of carbon compounds to CO_2 and H_2O. As an example, brewer's yeast may ferment glucose to ethyl alcohol and CO_2.

$$C_6H_{12}O_6 \rightarrow 2C_2H_5OH + 2CO_2; \quad \Delta F = -55 \text{ kg-cal/mol} \qquad (VII\text{-}4)$$

There is no participation of atmospheric oxygen in this reaction, and in fact many organisms are obligatory anaerobes, that is, they can grow only in the absence of oxygen. The question may be raised, therefore, whether it was fair to use the combustion of glucose as a type reaction in discussing the overall energy balance-sheet. But when one kind of organism does not carry the oxidation of carbon compounds as far as CO_2 and H_2O, eventually other types complete the oxidation to this final stage. Moreover, those organisms which do not complete the oxidation of carbon compounds probably represent a small proportion of the total energy metabolism of living systems. Thus, for the rough overall purpose for which it was used, reaction VII-3 seems applicable enough.

The term biological oxidation is usually applied to reactions in which O_2 participates, such as VII-3, whereas those such as VII-4 are called fermentations.[13] But in modern usage the term oxidation need not imply the uptake of oxygen, and although alcoholic fermentation is not, strictly speaking, an oxidation, it involves certain oxidative steps. Essentially, oxidations of all kinds involve the transfer of electrons from the substance being *oxidized* to another substance which is simultaneously *reduced*. The removal of H atoms from a compound involves the removal of electrons just as does the addition of O_2; in either case the compound is oxidized. The predominant type of oxidation going on in living systems is *dehydrogenation*, the removal of hydrogen atoms. For example, the oxidation of an aldehyde to an acid, an important step in some biological processes, may be written,

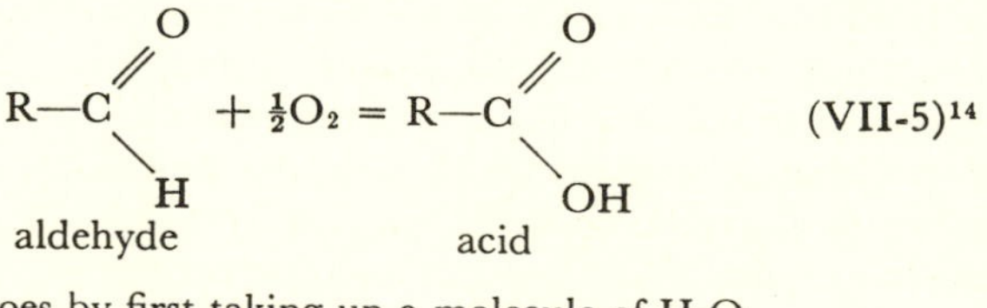

but it probably goes by first taking up a molecule of H_2O,

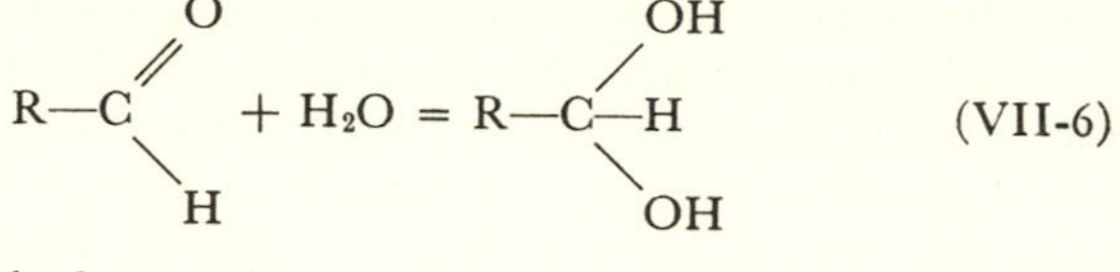

Then follows a dehydrogenation,

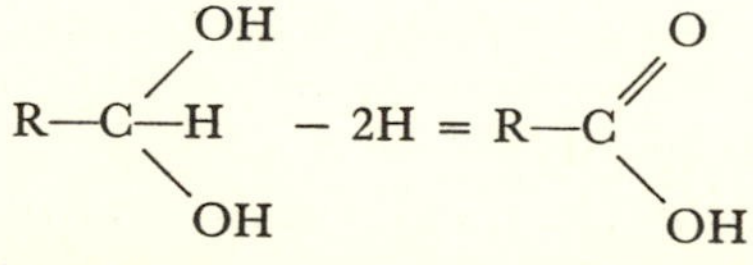

[13] David R. Goddard discusses the somewhat confused usage of terms in this field in *Science* (1945) 101, 252.

[14] The symbol R is used to indicate any attached grouping, for example, CH_3.

and the hydrogens combine with oxygen,

$$2H + \frac{1}{2}O_2 = H_2O$$

Both the oxidation and fermentation of glucose go, as will be seen, through various steps involving at one point the formation of pyruvic acid, so we may write as a part of both processes,

$$C_6H_{12}O_6 \rightarrow 2CH_3COCOOH + 4H \qquad \text{(VII-7)}$$

Up to this point both processes may be called dehydrogenations. In a condensed schematic form which will describe this and most oxidations in living systems, we may write,

$$AH_2 + B \rightarrow A + H_2B \qquad \text{(VII-8)}$$

where AH_2 is a *hydrogen donor*, and B is a *hydrogen acceptor*. In reaction VII-5, 6, aldehyde (the hydrogen donor) is oxidized to acid; O_2 (the hydrogen acceptor) is reduced to H_2O. In reaction VII-7 glucose is the hydrogen donor, and although the hydrogen acceptor is not specified in the equation it must appear somewhere in the process; as written this reaction is incomplete.

The fermentation of glucose to ethyl alcohol is perhaps understood more completely than any of the processes associated with biological oxidation, and will serve to illustrate certain important aspects of such processes. The scheme in Figure 13 summarizes what is known of this process, illustrating its stepwise nature. All the steps may be carried out by yeast "juice," which contains none of the living yeast cells but does contain the requisite enzymes, substances derivable only from living systems. Certain of the steps are of particular interest because they are common to many biological processes. Indeed, the whole scheme may represent a basic pattern included to a greater or less extent in all biological oxidations. For example, with modification of only the first and last steps, the scheme would represent the chemical changes associated with the anaerobic contraction of muscle. The legend of the figure gives the names of specific enzymes and coenzymes, substances without which the particular reaction steps would not proceed. The enzymes are all proteins, complex substances to be discussed in the next chapter; the coenzymes are substances of various types, less complex than the enzymes. In a general sense, the members of both groups may be regarded as catalysts. It is common to think of a catalyst as a substance which forwards a reaction without itself undergoing permanent alteration. The catalyst may be temporarily altered but returned to its original state before the end of the reaction. That is, it may be repeatedly altered and regenerated in a cyclical manner, and so only a very small quantity of the catalyst is required

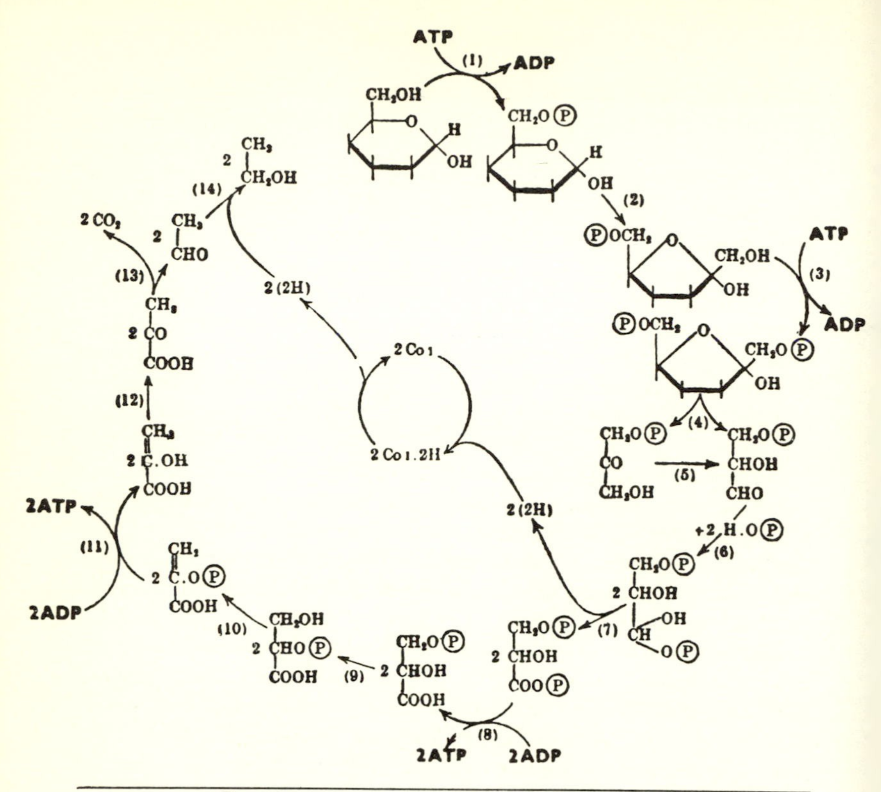

	Reaction step	Enzyme	Coenzyme
(1)	glucose → glucopyranose-6-phosphate	hexokinase	ATP
(2)	→ fructofuranose-6-phosphate	oxoisomerase	....
(3)	→ fructofuranose-1:6 diphosphate	phosphohexokinase	ATP
(4)	→ 3-phosphoglyceraldehyde and phosphodihydroxyacetone	zymohexase	....
(5)	phosphodihydroxyacetone → 3-phosphoglyceraldehyde	phosphotrioseisomerase	....
(6)	→ 1:3 diphosphoglyceraldehyde	?	H_2PO_3
(7)	→ 1:3 diphosphoglyceric acid	triosephosphate dehydrogenase	Co I
(8)	→ 3 phosphoglyceric acid	phosphokinase (unnamed)	ADP
(9)	→ 2 phosphoglyceric acid	phosphoglyceromutase	2,3-phosphoglyceric acid
(10)	→ phospho-enol-pyruvic acid	enolase	Mg ions
(11)	→ enol-pyruvic acid	phosphokinase (unnamed)	ADP
(12)	→ pyruvic acid	?	....
(13)	→ acetaldehyde + CO_2	carboxylase	cocarboxylase
(14)	→ ethyl alcohol	alcohol dehydrogenase	Co II

FIGURE 13. Summary of the reaction steps in the alcoholic fermentation of glucose by "yeast juice." The circled P = the phosphate group, (ph); ATP = adenosine triphosphate; ADP = adenosine diphosphate. (From E. Baldwin, *Dynamic Aspects of Biochemistry*, Cambridge (1948) Cambridge Univ. Press, by courtesy of the author and the publisher.)

to expedite the change of large amounts of reactants to products. This brief description of course leaves out the energetics of the process. But let us examine the cyclical role of some of the coenzymes in the above scheme before considering the thermodynamic and kinetic aspects.

At the central point of the fermentation scheme (Figure 13), like the hub of a wheel, is a cycle involving coenzyme I (Co I), which ties together reactions (7) and (14); this cycle is represented separately in Figure 14. Reaction (7) consists in the dehydrogenation of 1:3 diphosphoglyceraldehyde to 1:3 diphosphoglyceric acid. The two hydrogens lost by the former are taken up by Co I, which is thus

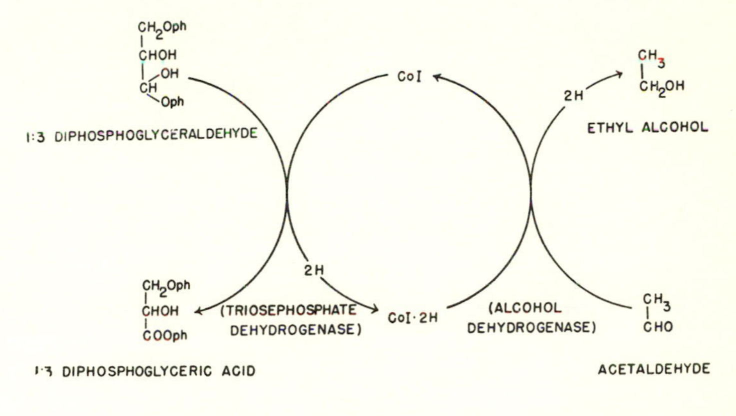

FIGURE 14

reduced to Co I.2H. This reaction requires the enzyme triosphosphate dehydrogenase and Co I. In reaction (14) Co I.2H gives up the 2H received from reaction (7) to acetaldehyde, which is thus reduced to ethyl alcohol; the coenzyme is returned to its oxidized form Co I. Coenzyme I may be repeatedly reduced and reoxidized, returning always to the same starting point; so that it is not itself used up. From the standpoint of the reaction represented in Figure 13 or that part of it represented in Figure 14, coenzyme I may be regarded as a catalyst in the usual sense, whereas it must be regarded as a component of reaction (7) or (14) if these are to be considered as separate entities. Obviously the term catalyst must be somewhat elastic when applied to a living system, since a substance such as coenzyme I, which is a reactant in a given step, may serve as a catalyst in the overall process. One is not, of course, interested so much in defining terms as in understanding mechanisms, and the coenzyme I cycle serves this purpose as an example which is paralleled elsewhere in living systems.

Included in the fermentation scheme is another system particularly important because of its apparently unique role in the mobilization and transfer of free energy within living systems. The symbols ATP and ADP, associated with reactions (1), (3), (8), and (11) in Figure 13, represent the compounds *adenosine triphosphate* and *adenosine diphosphate* respectively. These compounds are phosphates of *adenylic acid*, also known as *adenosine monophosphate*

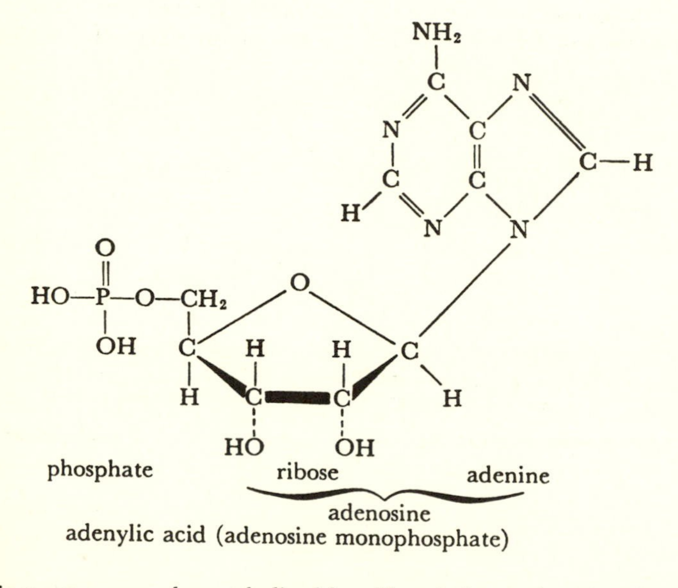

This structure may be symbolized by abbreviating to A—ph, where A represents the adenosine part of the molecule, and —ph the phosphate part. The addition of phosphate groups gives adenosine diphosphate, ADP,

$$\text{A—ph} \sim \text{ph}$$

and adenosine triphosphate, ATP,

$$\text{A—ph} \sim \text{ph} \sim \text{ph}.$$

The three compounds are sometimes referred to collectively as the adenylic acid system. In ADP and ATP, the symbol ~ph, represents an "*energy-rich phosphate bond*," to be contrasted with the "*energy-poor phosphate bond*," symbolized by —ph, which is characteristic of the adenylic acid part of these molecules.[15] The free energy concerned in reactions involving only the transfer of these types of bonds, may be estimated, and is commonly referred to as the "free energy of the energy-rich (or energy-poor) phosphate bond." For the energy poor phosphate bond, —ph, ΔF is about 2 kg-calories per mole; for the energy-rich phosphate bond, ~ph, ΔF ranges from about 10 to 12 kg-calories per mole, according to the molecular configuration in which it occurs.

In certain reactions, energy-rich phosphate bonds may be transferred from ATP to some other molecule, with increase in the free energy of that molecule. To illustrate, reaction (1) of the fermentation scheme may be written

$$\text{glucose} + \text{ATP} \rightarrow \text{glucopyranose-6-phosphate} + \text{ADP} \quad \text{(VII-9)}$$

or in two steps

$$\text{ATP} \rightarrow \text{ADP} + \sim\text{ph}$$

and

$$\text{glucose} + \sim\text{ph} \rightarrow \text{glucopyranose-6-phosphate}$$

The glucose has been *phosphorylated*, adding one ~ph to form glucopyranose-6-phosphate which has higher free energy than glucose. The phosphorylated compound may take part in reactions in which glucose cannot. The ATP by giving up one ~ph reverts to ADP. Here the ATP-ADP system can only take part in a continuous cycle if ATP is continually renewed by the addition of ~ph bonds from the outside; this means that free energy must be added to the system.

Under other circumstances we find ADP being reconverted to ATP, for example in reaction (11) of the fermentation scheme. We need first to consider reaction (10) of the scheme, in which the phosphate bond is prepared for transfer

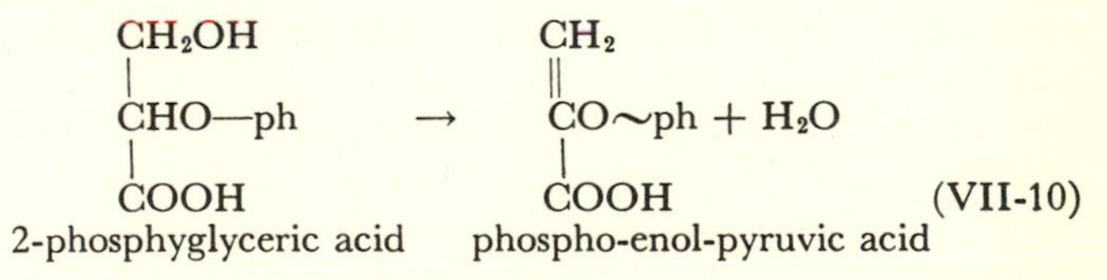

By the removal of one molecule of water, 2-phosphoglyceric acid becomes phospho-enol-pyruvic acid. The rearrangement of the molecular structure results in the "mobilization" of free energy in the region of the phosphate group, transforming the energy poor phosphate

[15] The terms "energy-rich phosphate bond," and "energy-poor phosphate bond" were introduced by Fritz Lipmann, who originated this concept.

bond into an energy-rich phosphate bond. Thus the molecule is prepared for the next step, reaction (11), in which the ~ph bond is transferred to ADP in forming enol-pyruvic acid and ATP.

$$\begin{array}{ccc} CH_2 & & CH_2 \\ \| & & \| \\ CO\sim ph + ADP & \rightarrow & C{-}OH + ATP \\ | & & | \\ COOH & & COOH \\ \text{Phospho-enol-pyruvic acid} & & \text{enol-pyruvic acid} \end{array} \qquad (VII\text{-}11)$$

In this step and also in reaction (8), free energy concentrated in the energy-rich phosphate bond is tapped off from the fermentation process. This free energy may be used elsewhere in living processes. Some of it may be returned to the fermentation process in reactions (1) and (3), where ~ph bonds from ATP are required for the forwarding of specific reaction steps.

If all the phosphate bond transfers in the fermentation scheme (Fig. 13) are counted, the score is, for every glucose molecule fermented; 2 ~ph removed from ATP to the reaction (one each in reactions 1 and 3); 4 ~ph added to ADP by the reaction (2 each in reactions 8 and 11); total 2 ~ph which can be transferred by ATP to some other process outside the cycle. For the overall reaction (VII-4), ΔF is about -50 kg-calories per mole, and the two energy-rich phosphate bonds derived therefrom represent a free energy change of about 20 kg-calories per mole which can be utilized in outside work, suggesting a maximum possible efficiency of around forty per cent. For the complete aerobic oxidation of glucose to CO_2 and H_2O, ΔF is -688 kilogram-calories per mole. This oxidation proceeds through a series of steps in which ATP is repeatedly involved, and it is estimated that about fifty ~ph bonds may be made available in its course, amounting to about 500 kilogram-calories per mole, so the maximum efficiency might be in the neighborhood of seventy per cent. The maximum overall efficiency for these processes may seem remarkably high, but there appears no evidence of violation of thermodynamic principles.[16]

The fermentation of glucose to ethyl alcohol has been chosen as an example because, although it does not represent the main stream of biological oxidation, it offers good illustrations of the stepwise nature of such processes. The oxidation of glucose to CO_2 and H_2O has been worked out in a comparable fashion, although some uncertainties

[16] It is important to keep in mind that one does not deal here with a heat engine, but an isothermal process in which the theoretical maximum efficiency in terms of free energy is 100 per cent.

remain. There are more steps, and there are enzymes and coenzymes involved that are not concerned in fermentation; but the same kind of energetic picture is presented. In both cases the reaction is broken up into a series of steps, each of which is "reversible" in that it does not involve a very great free energy change; none are greater than that of an energy-rich phosphate bond. On the other hand the overall reactions are "irreversible" in the sense that their high free energy changes prevent them from going in the opposite direction in a single step. But this does not mean that they could not be reversed if carried out in a series of successive steps, a point that will need comment later on. The metabolism of a variety of organisms has been studied in greater or less detail and although much exploration remains to be done a similarity among the pathways followed is already discernable, adding evidence of the common ancestry of living things to that already accumulated in other fields. It would be fascinating to trace these evolutionary relationships, but for this the reader must go to other sources.[17]

In muscular contraction free energy of biological oxidation is utilized to accomplish mechanical work, and here ~ph bonds play an intimate role in transferring the free energy to the contractile mechanism. Muscle may contract either in the presence or absence of oxygen. In anaerobic contraction, the reaction which yields the free energy is very like that of alcoholic fermentation, but starts from glycogen and goes through pyruvic acid to lactic acid instead of ethyl alcohol. In both cases the ATP-ADP system plays its important role of ~ph bond transfer. There is in muscle provision for the "storage" of ~ph bonds by transfer from ATP to the guanidine bases, creatine and arginine, to form *phosphagens*, the phosphates of these compounds. In the muscles of vertebrate animals phosphagen is formed from creatine

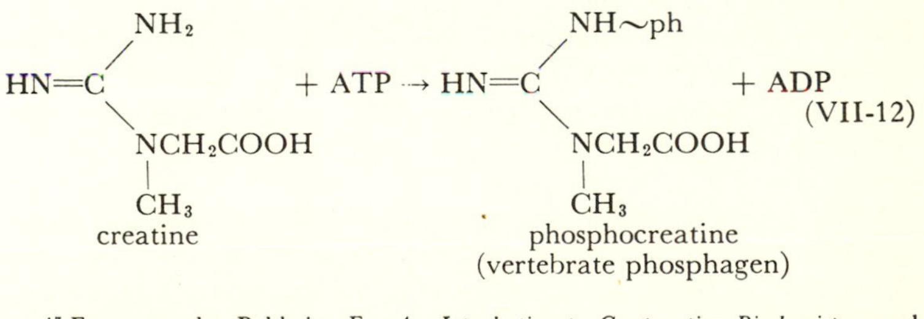

[17] For example, Baldwin, E., *An Introduction to Comparative Biochemistry*, and *Dynamic Aspects of Biochemistry;* Barron, E. S. G., *Mechanism of Carbohydrate Metabolism;* Florkin, M., *l'Evolution biochimique;* Foster, J. W., *The Chemical Activities of Fungi;* Kluyver, A. J., *The Chemical Activity of Microorganisms;* Lwoff, A., *l'Evolution physiologique.* For complete citations see Bibliography.

and comparably in the muscles of invertebrate animals a phosphagen is formed from arginine.

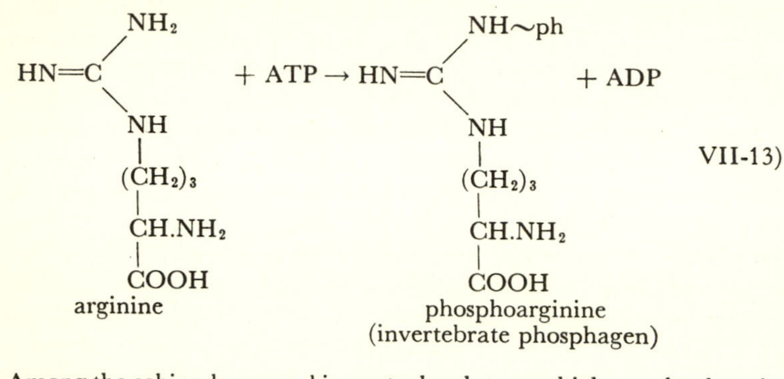

Among the echinoderms and in protochordates—which may be thought of evolutionarily as at a transition point from the invertebrates to the vertebrates—creatine and free arginine both appear, but elsewhere

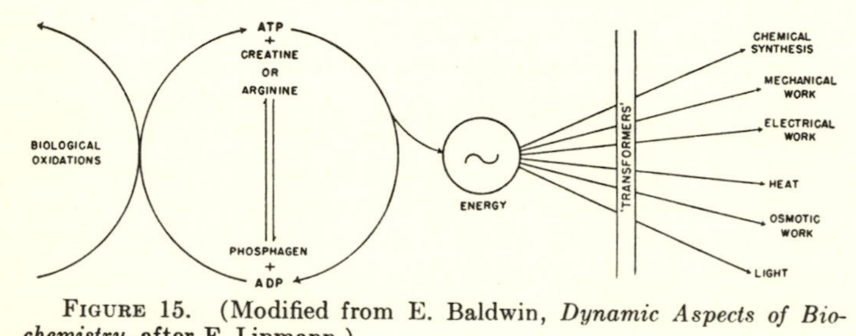

FIGURE 15. (Modified from E. Baldwin, *Dynamic Aspects of Biochemistry,* after F. Lipmann.)

they are not found side by side. The phosphagens constitute a reserve of ~ph bonds, from which free energy may be released to the contractile machinery by means of the ATP-ADP system. They are regenerated by transfer of ~ph bonds from ATP. In muscles under extreme depletion, the ~ph bond of ADP may be released with the formation of adenylic acid, but under ordinary conditions this does not occur.

The utilization of the energy from biological oxidation for the various living processes may be visualized as in Figure 15. Here the ATP-ADP system is seen as the central "wheel" in the energy transforming machine. The guanidine base-phosphagen system is shown in its secondary role in muscular contraction, within the ATP-ADP cycle. On the left, biological oxidation feeds free energy into the ATP-ADP system in the form of ~ph bonds which are transferred to ADP to form ATP. From the ATP-ADP "wheel" the free energy of ~ph bonds from ATP is introduced through proper transforming mechanisms into the various biological processes represented on the right; the synthesis of the chemical compounds required for growth, maintenance, and reproduction; mechanical work accomplished by muscular contraction; etc. There is much left to be learned about various aspects of free energy transformation in these later stages, but the fundamental role of the energy-rich phosphate bonds and the ATP-ADP system seem clearly established.

The unique role played by the ∼ph bond would seem to imply a similarly unique role for the element phosphorus. Why, one may ask, does this element, in a particular combination with hydrogen and oxygen, lend itself to the transfer of free energy from place to place in living systems? Why only this element, and not nitrogen, its congener in the periodic table, which is similar in its properties? Does the unique behavior of phosphorus in this case depend upon very special properties, such that even quite similar elements could not serve as substitutes; or is this unique function to be attributed to events in evolutionary history which caused the accidental "selection" of phosphorus for this rôle? A possible answer to this particular question will be offered in a later chapter. The problem is a general one, however, which has already been posed in discussing the fitness of the environment.

As in the case of coenzyme I, cited above, the ATP-ADP system may be thought of as a catalyst with reference to the overall reaction; this is obvious from the model in Figure 15, which shows the latter as operating in a cycle including the whole metabolic process. This point of view may not apply, however, to isolated reaction steps. In most of the steps in the fermentation scheme, specific enzymes as well as coenzymes are required. It is more difficult to show that enzymes act in a cyclical fashion, but they are able to catalyze the reaction of a large amount of the specific substance—the *substrate*—upon which they act. The enzymes are more specific in their action than the coenzymes, acting only upon very closely related substrates in most cases, and sometimes upon only a single compound. Some enzymes require a coenzyme, others do not. The specificity of the enzymes may depend upon the spatial characteristics of their molecules, a matter that will be discussed in the next chapter. It seems probable that there is always some degree of combination between the enzyme and its substrate, and between the enzyme and the coenzyme, but the closeness of this union may vary. In some instances it may be a matter of definition whether one considers a substance to be a coenzyme, or merely a more or less nonspecific adjuvant.

In the diagram of energy transformation in Figure 15, free energy is represented as being utilized for chemical synthesis; that is, the production of compounds for the building and operation of living organisms, for their growth, maintenance, and reproduction. Such processes will be largely the subject of the next chapter, but at this point a discussion of the energetics of synthesis may be introduced, since the thermodynamics of biological oxidations, enzyme catalysis, and the distribution of free energy, are inextricably bound together. The equilibrium state of a chemical reaction may be represented as follows,

$$\frac{[\text{products}]}{[\text{reactants}] \times \text{equilibrium constant}} = 1 \qquad \text{(VII-14)}$$

The concentrations of the reactants and products, indicated by placing these words in brackets, are determined by the equilibrium constant. If the concentrations of the reactants and products are other than those specified by the above relationship, the reaction will tend to go in the direction of equilibrium: if the concentration of reactants is increased more products form; if the concentration of the products is increased more reactants form. If the equilibrium constant has a low value, it is possible to push the reaction in one direction or the other by the addition of reactants or products—the process is readily reversible. If, on the other hand, the equilibrium constant has a high value, the addition of reactants will always result in more products, but the addition of products will have very little effect in increasing the concentration of reactants. Catalysts speed up the attainment of equilibrium without altering the equilibrium itself. From a thermodynamic standpoint they do this by lowering the free energy required for activation. If the reaction is reversible the enzyme will speed it in one direction as readily as the other, according to the concentrations of reactants and products. Thus, enzymes are sometimes said to be "reversible" in their action, but, of course, this only applies in chemical reactions which are themselves readily reversible.

What has been said about equilibria and enzymes applies to a single reaction, but suppose that we deal with a series of successive reversible reactions each requiring catalysis, say a series of reactions in a living system, each with a specific enzyme. This may be indicated by

$$A \overset{\alpha}{\rightleftarrows} B \overset{\beta}{\rightleftarrows} C \overset{\gamma}{\rightleftarrows} D \qquad \text{(VII-15)}$$

where A, B, C, D, represent a series of reacting substances and α, β, γ, represent enzymes catalyzing the indicated reversible reactions. Let us suppose the free energy relationships to be such that at equilibrium all the substances, A, B, C, and D, are present in equal amounts. If now we add an excess of B, there must be a tendency to readjust toward equilibrium by forming all the other substances. But suppose enzyme α is present to catalyze the reaction $A \rightleftarrows B$, but β is not present to catalyze reaction $B \rightleftarrows C$. Then substance A must be formed until the concentrations of A and B are equal, but there will be no appreciable increase in C and D. Given sufficient time all the reactions would come to equilibrium, and all would have equal concentrations, but for any short period—and this is what counts in the living system—the concentrations of A and B will be higher than those of C and D. In an analogous fashion enzymes may regulate the concentrations of various substances in the cell so that conditions may be maintained which do not approach closely to equilibrium.

In all the discussion of reversal of reactions it should be noted that there has been interference from outside the reaction system itself. For example, the phrases, "if the concentration of products is increased," and "addition of reactants," indicate such interference. In an *in vitro* system, the interference is by the experimenter, who may add substances at will. The act of adding or removing substances involves doing work on the system and this work involves the expenditure of free energy. If it is the experimenter who does the work, the free energy comes from his own metabolism; but a comparable readjustment within the living system must require the expenditure of free energy from the metabolism of that system itself.[18] Thus the mobilization of the necessary reactants at the appropriate point in the cell requires the expenditure of free energy, even though the reaction in which they participate is itself exergonic, and may indeed be the source of energy for the mobilization process.[19]

The important role of the ATP-ADP system in the mobilization and transfer of free energy has already been discussed, and it may serve again to illustrate the difficulty of applying the terms exergonic and endergonic to parts of the living mechanism. In reaction VII-11, which is reaction (11) of the fermentation scheme (Fig. 13), ATP is formed by transferring to ADP a $\sim$ph bond from phospho-enol-pyruvic

[18] As a useful figment, James Clerk Maxwell invented his famous "demon" which could direct the course of molecules and so might cause a reaction to go in the direction of less entropy. From time to time it is suggested that living systems, since they cause a local decrease in entropy, possess, or are possessed of, a Maxwellian demon. This idea may stem from the failure to realize that the demon, like the experimenter, must have a metabolism, a matter which seems clear enough. Since the efficiency of the living system is such as to allow for the metabolism of the demon, it seems that we need not be too much concerned about him in our thermodynamic considerations. After reading this footnote, Walter Kauzmann called to my attention a formal treatment of this question by Leo Szilard, *Ztschr. f. Physik* (1929) *53*, 840–856.

[19] The efficiency of the overall process must, of course, be less than 100 per cent.

acid. Here one reaction is coupled with another with the result that one energy-rich compound is formed at the expense of another energy-rich compound. In the course of the reaction, there is a redistribution of free energy among the participating compounds, but the inclusive reaction is exergonic; it is catalyzed by an enzyme, phosphokinase, which presumably lowers the free energy of activation in some way. Left to itself with a limited amount of phospho-enol-pyruvic acid and ADP, the reaction would go to equilibrium and then stop. Only when it is possible continually to supply ~ph bonds from outside this particular reaction does the energy-rich ATP continue to be formed. This means bringing in free energy from the outside, and when the source of this free energy is included, the overall process is also exergonic. Only by restricting our view to very limited parts of the whole process can we designate any of the reactions as endergonic. In an overall sense the ATP-ADP system may be regarded as behaving as a catalyst which lowers the free energy of activation by acting in a cyclical manner.

In all these discussions of enzymes and reaction steps, single reactions have been pictured as parts of the living system as a whole. This picture has been based upon experiments in which parts have been removed from the whole—a procedure which is obviously necessary for such analysis, but which brings us perhaps not too close to reality when the reconstruction of the whole is essayed. One needs to think of the metabolism of living systems as the integration of reactions which are, separately considered, either exergonic or endergonic, but in the integrated process are strictly exergonic. To gain a general picture of these energy relationships, let us have recourse to an analogy effectively employed by E. S. G. Barron, which again makes use of the flow of a river. The combustion of glucose at high temperature may be thought of as comparable to the plunge of water over Niagara. In contrast one may think of the biological oxidation of this same compound as the safe passage of ships through the locks of the Welland Canal, between levels above and below the falls. In both cases the actual change in potential, as represented by the difference in level at the beginning and end of the canal, are the same. The passage of a ship through the canal is accomplished by a series of reversible steps, provided by the locks, in which it may be either raised or lowered. Whichever way the ship travels, however, potential energy of the river is expended. The direct combustion of glucose is highly irreversible, just as is the passage of Niagara in a barrel, because the amount of energy must be expended in a single leap. In the biological oxidation of glucose each step involves a much smaller free energy change, and many of them are readily reversible, in the same sense as passage through a canal lock. Yet the reversal of any of these steps must involve an expenditure of free energy in one way or another. In the long run the free energy change will be the same for the combustion of a given amount of glucose no matter how the process is carried out, just as the change in gravitational potential is the same for a given mass of water whether it flows over Niagara or through the locks of the Welland Canal. In either case, however, what is accomplished in the course of the flow of a given amount of potential depends upon the way the process is carried out.

In a living system true equilibrium is seldom if ever reached, but a *dynamic steady state* is maintained by the flow of energy through the system. In such a system reactions may take place which would not be expected if the reactants and products were mixed in a test tube, but there is really nothing in this which constitutes an escape from the second law of thermodynamics. Some of the reactions that go on in living systems are difficult or altogether infeasible to duplicate in the laboratory, under strictly *in vitro* conditions; the fact that some of the steps must go endergonically only adds to the difficulty. One such reaction of special interest is the formation of the peptide linkages in protein molecules, which will be discussed in the next chapter when the duplication of molecular patterns essential to living organisms is taken up. First, however, the thermodynamic picture should be completed by considering those truly endergonic processes that permit living systems to capture free energy from the outside world.

The Capture of Free Energy

Green plant photosynthesis is the principal means by which living systems capture energy from the outside, and the energetics of this process may be discussed before considering some others. Photosynthesis is essentially a photochemical reaction, and while it has never been carried out *in vitro*, it may be compared with photochemical reactions which have. The first step in any photochemical reaction is the absorption of radiant energy,

$$M + h\nu \rightarrow M^* \qquad \text{(VII-16)}$$

where M is a molecule of some substance, called the *light absorber*, $h\nu$ is a quantum of radiant energy, and M^* is the molecule M raised to an *activated state*. After this primary act, the activated molecule M^* may participate in chemical reactions in which the molecule M could not. The kind of reaction that occurs depends upon the nature of M and of the other molecules in its environment, for example

$$M^* + L \rightarrow ML \qquad \text{(VII-17)}$$

might represent such a reaction, where L is some molecule in the environment which can combine with M* but not with M. The energy of the quantum may in some instances supply energy to drive reactions endergonically. This is the case in photosynthesis, where the free energy of the products is greater than that of the reactants. The wavelengths that bring about typical green plant photosynthesis are those that are absorbed by the pigment *chlorophyll*, indicating that this substance is the light absorber. Thus we may write for the primary act in photosynthesis

$$\mathrm{Chl} + h\nu \rightarrow \mathrm{Chl}^* \qquad \text{(VII-18)}$$

where Chl is the chlorophyll molecule. There are several chlorophylls, all very similar structurally and all containing magnesium. The structure of a chlorophyll is shown in Table 9. These pigments are generally found in cell inclusions called chloroplasts. Chlorophyll is no doubt the principal light absorber in photosynthesis, and is commonly regarded as the only one. A closely related pigment, bacteriochlorophyll, serves as the light absorbing substance in the photosynthetic bacteria, and recent evidence[20] indicates that among diatoms and algae quite different substances, fucoxanthin, phycocyanin and phycoerythrin, may play this role. Reaction VII-18 may be used as a type, however, to represent the first step in photosynthesis. In what way chlorophyll participates in the photosynthetic reaction after absorbing the quantum of radiation is not clearly understood; it is not used up but apparently acts in some sort of "cycle."

The steps that follow this primary event are known only in a rather general way. To represent the overall reaction of green plant photosynthesis for thermodynamic purposes we may choose the reduction of CO_2 to glucose, writing it as below

$$\begin{aligned} CO_2 + H_2O &= 1/6(C_6H_{12}O_6) + O_2 \\ \Delta F &= +115 \text{ kg-cal/mol} \\ \Delta H &= +112 \text{ kg-cal/mol} \qquad \text{(VII-19)} \end{aligned}$$

This reaction is one-sixth of that written to represent the oxidation of glucose (VII-3), but going in the opposite direction; hence the energy values are one-sixth those given there, and have opposite sign. We might write instead

$$\begin{aligned} CO_2 + H_2O &= \{CH_2O\} + O_2 \\ \Delta F &\cong +120 \text{ kg-cal/mol} \\ \Delta H &\cong +112 \text{ kg-cal/mol} \qquad \text{(VII-20)} \end{aligned}$$

[20] Dutton, H. J., and Manning, W. M., *Am. J. Botany* (1941) *28*, 516; Emerson, R., and Lewis, C. M., *J. Gen. Physiol.* (1942) *25*, 579; Wassink, E. C., and Kersten, J. A. H., *Enzymologia* (1946) *12*, 3; and Haxo, F. T., and Blinks, L. R., *J. Gen. Physiol.* (1950) *33*, 389.

in which $\{CH_2O\}$ is a hypothetical compound representing the smallest unit of carbohydrate. The carbohydrates are compounds of C, H, and O, with the last two elements in the proportion H_2O; they include glucose and other sugars, starch, cellulose, and glycogen (animal starch). Such compounds may be considered as the principal products of photosynthesis. Whether glucose or another carbohydrate is considered as the end product, the values of ΔF, and of ΔH for the reduction of CO_2 are nearly the same, so there seems no objection to accepting reaction VII-19 or VII-20, including the energy values, as typical, thermodynamically, of the overall photosynthetic process. The value of ΔF in the latter reaction is slightly higher than in the former because the low partial pressure of CO_2 in the atmosphere has been taken into account, but for our purposes this difference is negligible. The symbol $\{CH_2O\}$ is used in VII-20 to distinguish the unit of carbohydrate from formaldehyde, CH_2O, which, although experimental evidence is now conclusively to the contrary, was long thought of as a possible product of photosynthesis.

Under any circumstances, the schematic nature of such reactions should be kept in mind, for although they are useful and necessary figments for overall thermodynamic calculations, they do not represent the true course of the process, which no doubt includes intermediate steps. But let us for the moment take reaction VII-20 literally, and imagine that one molecule of CO_2 is directly reduced to a unit of carbohydrate $\{CH_2O\}$, the necessary energy being supplied by the radiant energy absorbed by the chlorophyll. Radiant energy may be regarded as composed of discrete packets or *quanta*, which vary in size inversely with the wavelength of the radiation,

$$q = h\nu = h\frac{c}{\lambda} \qquad \text{(VII-21)}$$

where q is the energy of the quantum, h is Planck's constant, c the velocity of light, ν the frequency, and λ the wavelength.[21] Red light of wavelength as long as 0.68 μ may bring about green plant photosynthesis, and this corresponds to a quantum of 2.9×10^{-12} ergs of energy. To compare this value with ΔH and ΔF, it must be multiplied by the number of molecules in a mole (6.0227×10^{23}), and by the proper factor to convert ergs to kilogram-calories (0.239×10^{-10}). The value obtained is about 40 kilogram-calories per mole, only a little over one-third of the total energy change in reaction VII-20, for which ΔH is 112 kilogram-calories per mole. Thus, in the reduction of one molecule of CO_2 to carbohydrate, energy equivalent to at least three quanta of this size is needed, and this poses certain questions.

[21] $h = 6.6236 \times 10^{-27}$ erg-seconds; $c = 3.0 \times 10^{10}$ centimeters per second.

Of course the thermodynamic requirement of three quanta need not be closely related to the number actually used, but is a minimum set by the nature of the overall reaction. Under optimum conditions in the laboratory the net efficiency may be quite high, as few as four quanta being reported to be adequate.[22] Since some energy is no doubt required for energy of activation, and this is not recoverable, the use of only four quanta in carrying out a reaction requiring energy equivalent to at least three represents a surprisingly high efficiency. The gross efficiency under natural conditions is, as we have seen, much lower.

From the standpoint of thermodynamics there is no objection to the participation of three or four quanta in the reduction of a molecule of CO_2, but photochemically such a process occurring *in vitro* would be quite unique. That is, if chlorophyll captured three quanta and then passed on the whole of this energy to a CO_2 molecule, the process would have no parallel elsewhere in photochemistry. Ordinarily a simple inorganic molecule can hold a quantum it has captured for only a very short time, say 10^{-8} to 10^{-6} seconds, before the energy is lost by reemission, by collision, or in other ways; and it is only during this short period that the activated molecule may participate in photochemical reaction. Some organic molecules of complexity comparable to chlorophyll may, when activated by capture of a quantum, enter states in which they remain reactive for a considerably longer time; but in any case, the successive capture of three quanta and the summing of their energies within a molecule would be photochemically unique. In reality this difficulty only appears, however, when reaction schemes such as VII-19 and VII-20 are taken too literally. Photosynthesis is not a simple straightforward photochemical reaction, since it includes processes which go in the absence of light, some, at least, catalyzed by enzymes. This alone indicates the existence of intermediate steps, and if there were some way of accumulating energy in the course of these steps the energetic problem might be less troublesome.

Our point of view regarding the uniqueness of photosynthesis has shifted within recent years as the result of the comparative approach. It now appears that green plant photosynthesis may be best regarded as a special example among more general processes characteristic of living systems. This perspective on the problem we owe to C. B. van Niel, who has been occupied for a number of years with the chemistry of the autotrophic bacteria. The term autotrophic is applied to all those organisms that are able to manufacture their own basic supply of

[22] Only Warburg and Burk have found such efficiencies. A number of other investigators have found around 10 quanta to be required per CO_2 reduced, and such values are generally accepted.

energy-rich carbon compounds without other source of carbon than CO_2. In contrast the heterotrophic organisms must be supplied with energy-rich carbon compounds. In an overall sense the heterotrophs are dependent upon autotrophs for their energy supply, and as pointed out earlier, those autotrophs that carry on green plant photosynthesis account for much the greater part of the credit side of the energy balance-sheet; the other autotrophs interest us chiefly because of their evolutionary significance, and the possibility that they provide a key to the interpretation of autotrophic mechanisms in general. Presumably, all autotrophs reduce CO_2 to carbohydrate or related organic compounds, and are independent of outside sources of energy-rich carbon compounds, but they accomplish this through chemical reactions that appear at first glance to be quite different.

Two general groups of autotrophic bacteria may be recognized: the *photoautotrophic bacteria* which utilize the energy of sunlight, and the *chemoautotrophic bacteria* which are able to obtain their required free energy from exergonic inorganic reactions. The reduction of CO_2 by the photoautotrophic bacteria is generally referred to as *bacterial photosynthesis*, sometimes as photoreduction. The reduction of CO_2 by chemoautotrophic bacteria is commonly called *chemosynthesis*. Species of photoautotrophic bacteria may reduce CO_2 in a variety of ways when supplied with light energy, e. g. by the oxidation of H_2S to elementary sulfur, according to the overall reaction

$$CO_2 + 2H_2S = \{CH_2O\} + H_2O + 2S \qquad \text{(VII-22)}$$

Some use other inorganic sulfur compounds oxidizing them to sulphate, and some may also utilize a variety of organic compounds, e.g. isopropanol

$$CO_2 + 2CH_3CHOHCH_3 = \{CH_2O\} + H_2O + 2CH_3COCH_3 \qquad \text{(VII-23)}$$

Some species may oxidize hydrogen gas

$$CO_2 + 2H_2 = \{CH_2O\} + H_2O \qquad \text{(VII-24)}$$

All these reactions, like those written for green plant photosynthesis, must be regarded as overall reactions only, and not as indicating the mechanism by which they proceed. While they appear superficially quite different, van Niel has pointed out that all may be described by the scheme

$$CO_2 + 2H_2A = \{CH_2O\} + H_2O + 2A \qquad \text{(VII-25)}$$

where H_2A represents a hydrogen donor which is oxidized to A. In reactions VII-22 to VII-24, the application of this equation is at once obvious. In green plant photosynthesis H_2O may be regarded as the hydrogen donor, and the reaction written

$$CO_2 + 2H_2O = \{CH_2O\} + H_2O + O_2 \qquad (VII\text{-}26)$$

which is thermodynamically equivalent to VII-20, since one molecule of H_2O on each side of the equation cancels out. The implication that water is oxidized in green plant photosynthesis has received experimental support from experiments with isotopic tracers. By introducing a heavy isotope of oxygen (O^{18}) into either the CO_2 or H_2O taking part in photosynthesis, it has been shown that the O_2 produced comes from the latter, not, as the formal reaction seems to require, from the former. It has also been shown by R. Hill, and subsequently by others, that suspensions of isolated chloroplasts in the presence of a suitable oxidizing agent can, when exposed to light, bring about the oxidation of H_2O with the production of O_2; no CO_2 is reduced, so this process represents only a part of photosynthesis.

The chemoautotrophic bacteria carry out the reduction of CO_2 by utilizing the energy from a variety of exergonic inorganic reactions including the oxidation by O_2 of nitrites, sulfides, sulfur, hydrogen, etc. It is of greatest interest to find that all the photosynthetic reactions carried out by the photoautotrophic bacteria with the aid of light in the absence of O_2 may, by some species of these bacteria, be carried on in darkness, but only if O_2 is then present. Thus, in a sense, light and O_2 seem to be interchangeable in these reactions. Another important finding, by Hans Gaffron, is that if the green alga *Scenedesmus* is adapted to anaerobic conditions it can utilize H_2 in the photosynthetic reduction of CO_2, whereas normally it carries on the green plant type of photosynthesis. Thus under one set of conditions this alga is able to carry out the reduction of CO_2 by a reaction typical of the photosynthetic bacteria, under other conditions by the reaction typical of the green plants. Moreover, *Scenedesmus* may be made to carry out the reduction of CO_2 in the dark in the presence of O_2, just as do the photosynthetic and some of the chemosynthetic bacteria. Here is seen ample evidence of an interrelation between all the types of photosynthesis and of chemosynthesis. If an organism like *Scenedesmus* can be made to carry on all these types of autotrophic activity, it seems reasonable to conclude that all the autotrophic forms possess similar basic mechanisms for the reduction of CO_2 to carbohydrate or other comparable organic material. We see here evidence for a common ancestry among all the autotrophs.

It seems that we may extend this biochemical relationship even further, to include organisms ordinarily considered to be strictly heterotrophic. It has been demonstrated within recent years that many kinds of plant and animal cells are able to assimilate CO_2, but probably without completely reducing this compound to carbohydrate. Thus the "fixation" of CO_2, which might be thought of as a first step in autotrophic metabolism can be carried out by heterotrophs as well as autotrophs. This process may represent the assimilation of CO_2 into the carboxyl structure

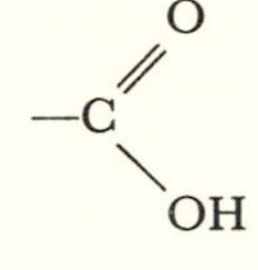

characteristic of organic acids, which would not itself involve a large free energy change, and hence does not answer the energetic problem of CO_2 reduction to carbohydrate, although it may represent an intermediate step. It does, however, indicate a general interrelationship between the autotrophs and the heterotrophs, making our categorical separation of the two types of living system perhaps more arbitrary than real. It also indicates that the search for the point of impingement of the photochemical component of photosynthesis should be sought elsewhere than the CO_2 molecule.

We may return now to the question of the summing of several quanta in photosynthesis. Is there some single photochemical step which can account for the capture and temporary storage of a quantum of energy, four of these steps serving to capture all the energy required for the reduction of a CO_2 molecule to carbohydrate? We have seen evidence that the crucial step in green plant photosynthesis may be represented as the oxidation of water, and we might think of the water molecule as "split," as a first act in photosynthesis. The energetics of this splitting is difficult to evaluate, however, without knowing more about the reaction, so it is uncertain whether this would help us out of our difficulty. It may be pointed out that a common basis for bacterial and green plant photosynthesis is to be expected, and that the overall energies are less for some of the former than for the latter. For example the ΔH for reaction VII-22 is only about 22 kilogram-calories per mole. Again, the purple sulfur bacteria carry out photosynthesis with wavelengths as long as 0.95 μ which corresponds to a quantum of about 30 kilogram-calories per mole,[23] as compared to the 40 kilogram-calories per mole that represent the smallest quanta that are known to forward green plant photosynthesis. It should be pointed out that in both cases the long wavelength limit may be set by the absorption spectrum of the light absorber—chlorophyll in the green plants and bacteriochlorophyll in the purple sulfur bacteria—rather than by the size

[23] French, C. S., *J. Gen. Physiol.* (1937) *21*, 71–87.

of the quantum, although the quantum efficiency may fall off with increasing wavelength. All these considerations throw doubt on the idea that the energy must always, of necessity, be introduced into the photosynthetic process in an amount equivalent to that of the quanta that are actually used. When we consider that chemosynthesis seems to be closely related to photosynthesis, and that in the former the energy supplied by an exergonic thermal reaction may be much less, mole for mole, than the energy of the quantum used in photosynthesis, we may have further reason to question the idea that the first step involves obligatorily the introduction of a minimum amount of energy as great as a quantum of red light.

It has been suggested that the energy absorbed as light quanta is stored temporarily in energy-rich phosphate bonds, and there is indeed some evidence for the participation of phosphate compounds in the photosynthetic process.[24] This idea is attractive, since we have already seen the unique function and general importance of these bonds, and of the ATP-ADP system, in biological oxidation. Could not the reverse of biological oxidation take place through the same or similar steps? If none of these steps required more energy than could be supplied by a ~ph bond, they might constitute a reversible reaction system which could be followed through with the high efficiency demonstrable in photosynthesis. It does not seem unlikely that the same or similar steps might be followed in the "uphill" photosynthetic and chemosynthetic processes as those that are followed in the "downhill" process of biological oxidation. Indeed this appears probable on the basis of the reasoning of comparative biochemistry. But we are met at this point by another problem. How can the energy of a light quantum representing, say, 40 kilogram-calories per mole, be divided among ~ph bonds representing about 10 kilogram-calories per mole, with relatively little loss? Assuming the maximum efficiency claimed, at least three ~ph bonds must be formed per quantum of red light used in green plant photosynthesis. Such a problem may seem as great, when regarded in terms of ordinary photochemical processes, as the alternative, reduction of the CO_2 molecule in a single step. But it is to be remembered that we do not deal with a simple system such as those in which photochemical reactions are ordinarily studied, but with an organized system of great complexity as compared to *in vitro* systems. One would not have guessed from a consideration of the overall reaction for the oxidation of glucose that this process could be carried out in a series of reversible steps, nor the complexity of these steps. But this is not the place to speculate about these matters, and it would be dangerous to do so without a much more complete examination of all the evi-

[24] Umbreit, W. W., Problems of autotrophy, *Bact. Rev.* (1947) *11*, 157.

dence. The reader must, therefore, be referred to more special literature on the subject, particularly the papers of van Niel.

No matter how carefully we examine the energetics of living systems, we find no evidence of defeat of thermodynamic principles, but we do encounter a degree of complexity not witnessed in the nonliving world. As compared to the *in vitro* photochemical and autoxidative reactions with which the chemist is more familiar, the complexity of autotrophic processes seems obvious, as is also the complexity of the step reactions in biological oxidations compared to the direct combustion of the same substances. To be sure, it is altogether probable that as investigation continues a relatively simple theme will be found connecting all the more or less isolated facts regarding energy metabolism in living systems, and indeed we have already evidence of that theme. But it seems certain that, simple or not in a general sense, quite complex molecules are involved. To be convinced of this one need only recall the role of enzymes in both the expenditure and accumulation of free energy. Most of the steps in biological oxidation require these substances, and their presence is also obligatory for CO_2 reduction, whether by photosynthesis or chemosynthesis. The enzymes themselves are highly complex molecules, the specificity of their action being apparently associated with this complexity. But such molecules, and the reproduction of their complex patterns, is a subject to be taken up in the next chapter.

The study of intermediate metabolism and photosynthesis has advanced greatly in recent years, but since accounts are available in biochemistry texts there seems little reason to review them here. Perhaps such names as H. A. Krebs and Melvin Calvin should at least be mentioned, however, in this connection. The examples and models used above may still suffice to illustrate the general energetic relationships with which we are concerned.

VIII · STRUCTURE AND ITS REPRODUCTION

"No structures are formed, as a rule, from spherical particles, and you will find it rather difficult to build any mechanism out of marbles."—A. SZENT-GYÖRGYI

STRUCTURE and reproduction are inseparable characteristics of living organisms; the understanding of one is contingent upon understanding the other. Essentially, the structure of any cell is a characteristic arrangement of specific kinds of molecules in interrelated patterns. Reproduction involves the duplication of the molecules themselves, and in addition the duplication of their arrangement in the proper pattern. The atoms composing the molecules are held together by forces we call chemical bonds, and the molecules are themselves held together in their characteristic pattern by other forces or bonds. So in attempting to explain structure and the manner of its reproduction, attention may be focussed on these bonds. The forming, breaking, and reforming of such bonds in the process of reproducing a living system or any of its parts—whether, for example, a cell, a virus, or a chromosome—should be susceptible to treatment in the same terms as any other chemical reaction; and one possible approach should be along the "broad highway of thermodynamics," roughly surveyed in Chapter III. Before proceeding to a consideration of structure and reproduction in living systems, then, it may be well to amplify what has already been said about chemical bonds.

Bonds between Atoms and Molecules

Knowledge of the forces, or bonds, which hold atoms and molecules together in nonliving systems has considerably increased in recent years, but a complete explanation of all the forces that maintain the structure of living organisms is still beyond our grasp. There can be little doubt, however, that the same kinds of chemical bonds are to be found in both living and nonliving systems; the only question that may reasonably be asked is whether further study will reveal the existence of hitherto unknown kinds of bonds, but there seems up to now no real necessity for assuming that such exist.

Chemical bonds include two important types which have already been discussed: the *covalent bond* in which electrons are shared by two atoms, and the *ionic bond* in which two charged atoms, or rather ions, of opposite charge are held together by their mutual attraction. Rarely are bonds exclusively of one type, most often they partake of the nature of both, resonating between one form and the other. In organic compounds the bonds are usually more nearly covalent than ionic. When a molecule is formed from its atoms, energy is lost; that is, there is a loss in total energy as referred to the atoms. The bond, then, represents a more stable condition, and in order to separate the atoms the energy lost when the bond was formed must be supplied. Hence, the more energy that has been lost, the more stable is the molecule. Correspondingly, the strength of the bond may be measured in terms of energy, and this energy is measured as ΔE at 0°K for the formation of the bond.[1] Chemical bonds are, in general, strong bonds, that is, they involve considerable amounts of energy, ranging, roughly, from 10 to 100 kilogram-calories per mole.

The bonds that hold molecules together are much weaker than those holding together the atoms in the molecule, so it takes less energy to separate molecules than to separate the atoms of a molecule. All molecules exert mutual attraction which only becomes effective at quite short distances, being greatest when the atoms of adjacent molecules just "touch." If atoms come still closer together so that their electron orbits begin to overlap, they tend to repel each other. Thus there is an average distance between molecules characteristic of the atoms composing them, at which they have a stable relationship. The molecules of a gas at low pressure are widely separated and exert little attraction upon each other; in this case they behave as separate particles, and the gas laws apply quite rigidly. But when the gas is compressed, the molecules are brought closer together and the effect of mutual attraction increases, resulting in deviation from the gas laws. The forces of attraction between molecules are generally known as *Van der Waals forces* after the physicist who first postulated them. In the liquid state, where molecules are much closer together than in gases, the Van der Waals forces become still more effective, and even more so in the solid state, where they serve to hold the molecules in definite mutual spatial relationship. These bonds characteristically involve energies less than 10 kilogram-calories per mole. The hydrogen bond, which joins hydrogen to a few other kinds of atoms, has already been described. These bonds are more nearly comparable to chemical bonds than to Van der Waals forces, but they may serve as

[1] See equation III-9 and accompanying discussion.

intermolecular bonds. Hydrogen bonds are weak bonds representing energies of only 5 to 10 kilogram-calories per mole, as a rule.

A complete description of a chemical bond cannot be given in terms of energy alone, but must include a description of the relative positions of the atoms or molecules in space. Thus the bond has a length, that is, the distance between the two atoms connected by the bond, which is characteristic of the kinds of atoms involved. For example, the length of the carbon to carbon single bond is 1.54 A, the carbon to carbon double bond 1.33 A. The distances between atoms held by Van der Waals forces corresponds to the sum of the effective radii of the atoms concerned, ranging from 1.2 A for hydrogen to double this value or more for the heavier atoms. In a polyatomic molecule the atoms are held in certain spatial relationships of which the angle which the interatomic bonds make with each other is measurable. The angles between the bonds as well as their lengths determine the characteristic shape of the molecule; for example, the angles determine whether the atoms which compose the molecule lie in a relatively flat plane, as is the case with the benzene ring, or is definitely three dimensional. The shape of the molecules and the way in which they are connected together in a crystal determines the shape of the crystal itself. The three dimensional arrangement of the atoms in a portion of a complex molecule is suggested in Plate II.

Crystals and Polymers

Crystals may be separated into three general groups, *ionic*, *covalent*, and *molecular*. The first two are found among inorganic substances; crystalline organic compounds are generally of the molecular type. In ionic and covalent crystals—sodium chloride and quartz, for example—separate molecules cannot be distinguished. In the former Na^+ and Cl^- ions are held firmly together by the mutual attraction of their opposite charges in such a way that the whole crystal may be regarded as a molecule. When such a crystal is dissolved, it goes directly into sodium ions and chloride ions, there being virtually no sodium chloride molecules in the solution. The crystals of most organic and some inorganic compounds are not formed of ions, however, but of molecules held together weakly by Van der Waals forces, or hydrogen bonds. These are molecular crystals. When such a crystal dissolves, the weak intermolecular bonds are broken and a solution of molecules results. The energy required to dissociate the molecule, on the other hand,

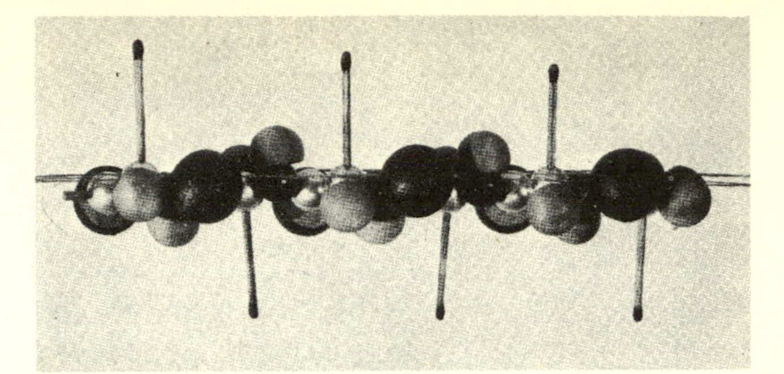

Plate I. Model of a short section of a protein "backbone" including five peptide linkages. The balls represent the molecules involved in these linkages; the match sticks represent the positions of the side chains. (From Neurath, H., *J. Phys. Chem.* (1940) *44*, 296, by courtesy of the author and the Williams and Wilkins Co.)

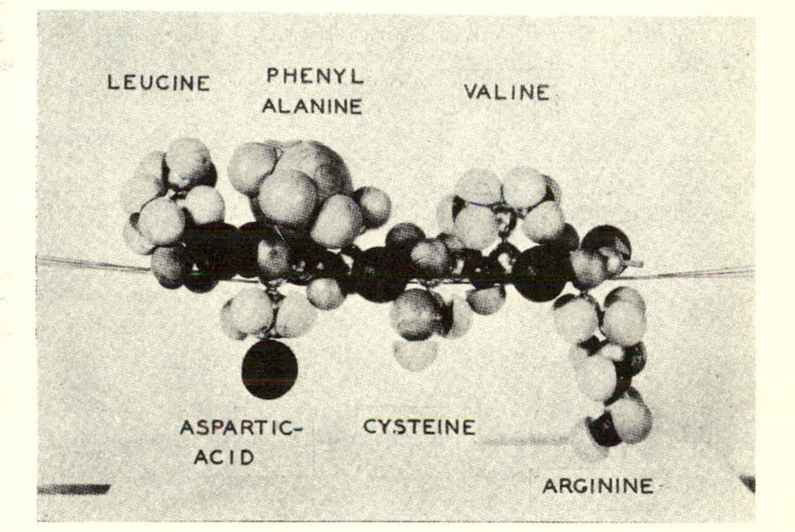

Plate II. The same as in Plate I, with models of amino acid groups replacing the match sticks. The diameters and volumes of the balls indicate the relative dimensions of the different molecules. (From Neurath, H., *J. Phys. Chem.* (1940) *44*, 296, by courtesy of the author and the Williams and Wilkins Co.)

involves chemical bonds and may be very high. Thus, the molecule in this kind of crystal is a stable entity which upon solution does not lose its character, and it may crystallize again by associating closely with other molecules in such a way as to be held by intermolecular forces.

The thermodynamics of the formation and solution of crystals needs brief mention here. A crystal forming from a supersaturated solution would seem to represent an increase in entropy if we include in our view only the crystal itself without regard to the solution in which it forms. As the solution cools it loses heat to the surroundings, and this loss would have to be included if we were to draw up a thermodynamic balance sheet.

True crystals are relatively rare in living material, and when they occur may be regarded as inclusions rather than parts of the living organism. All important, on the other hand, and somewhat analogous, are certain long chain-like *polymers*, made up of *monomer* units which are themselves molecular structures. *Proteins* and *nucleic acids* are such polymers, the monomers of which are, respectively, amino acids and nucleotides. The polymer chains may be coiled in characteristic helical fashion, being held in these spatial relationships by hydrogen and other weak bonds. An important difference between these polymers in living systems and simple crystals is a thermodynamic one, since energy must be supplied and work done in putting the polymers together.

Proteins

The proteins are important structural components of living systems, and play indispensible roles as catalysts—the *enzymes*—which serve to regulate rates of reaction and so determine the dynamic balance characteristic of living systems. An essential component of the protein molecule is the polypeptide chain, built up by the linkage together of numbers of amino acids. It is incorrect, however, to regard proteins merely as large polypeptides. Some twenty-odd different amino acids have been isolated from proteins, the structural formulas of nineteen of which are given in Table 6. Only five elements, C, H, O, N, and S, are present in these amino acids, although proteins may have attached groups which contain a few others. Molecular weights of proteins range at least as high as 6,760,000, which corresponds to about 50,000 amino acid groups. The joining together of the amino acids by the *peptide linkage* or *peptide*

Table 6. Amino Acids Which Have Been Isolated from Protein. All are α-amino acids and *l* forms.

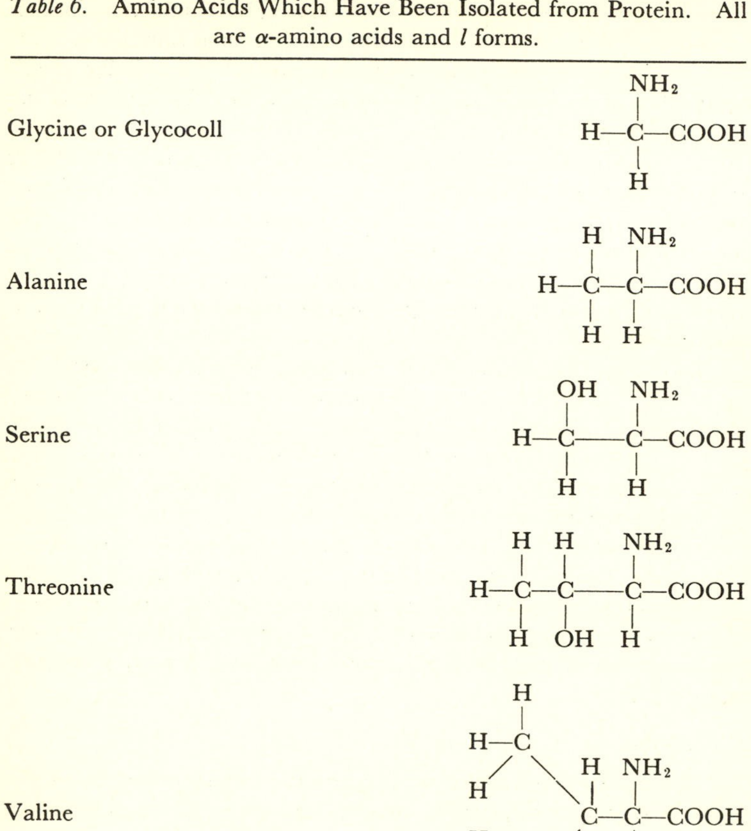

Table 6. (Continued)

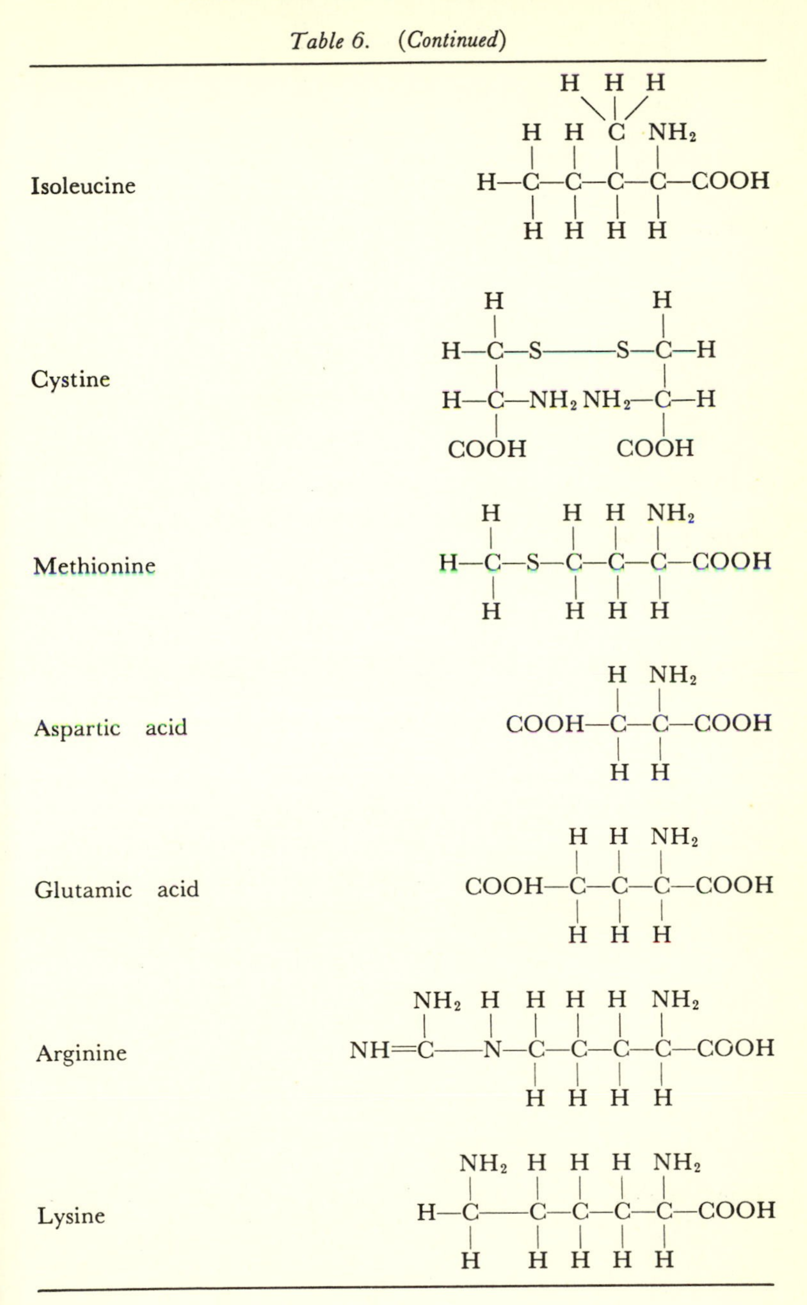

Table 6. (*Continued*)

Proline	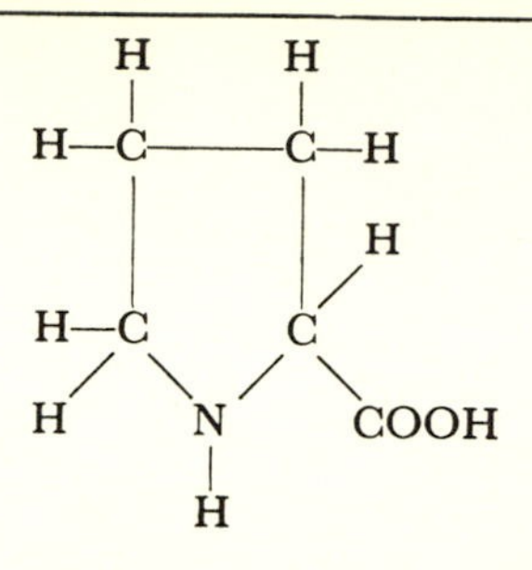
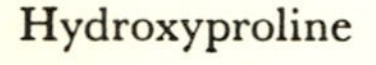 Hydroxyproline	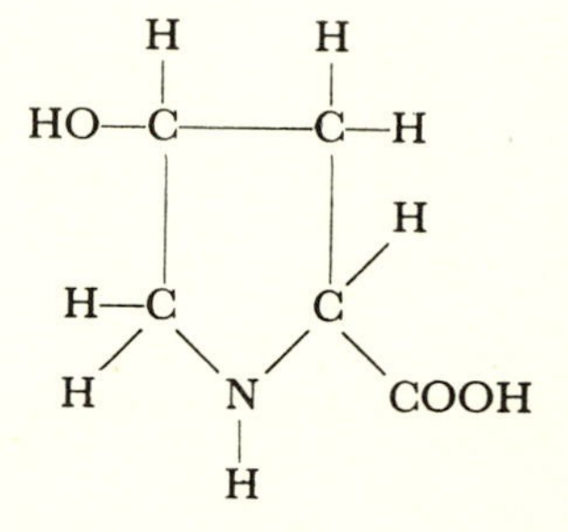
Tyrosine	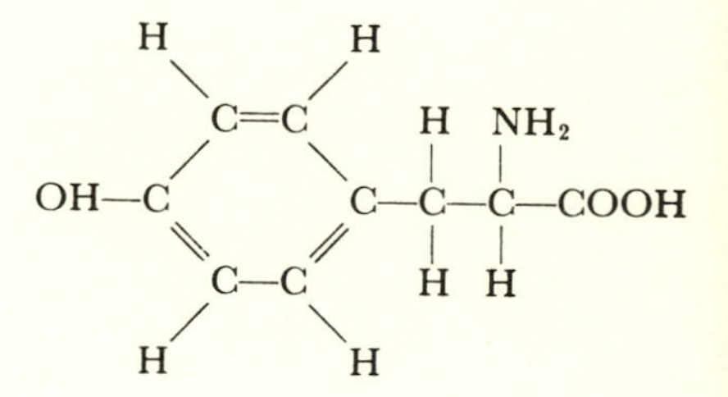
Phenyl alanine	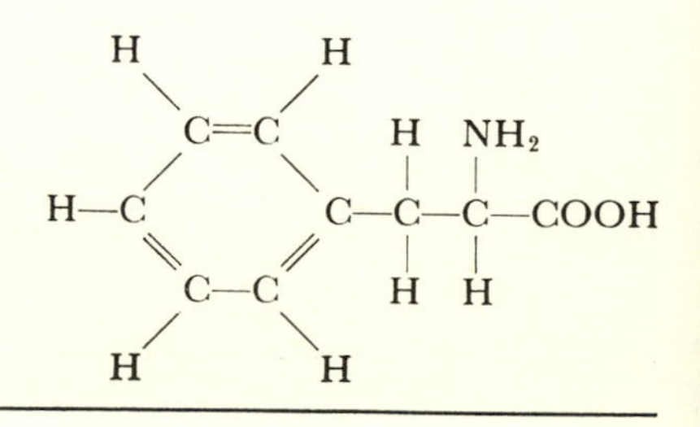

Table 6. (Continued)

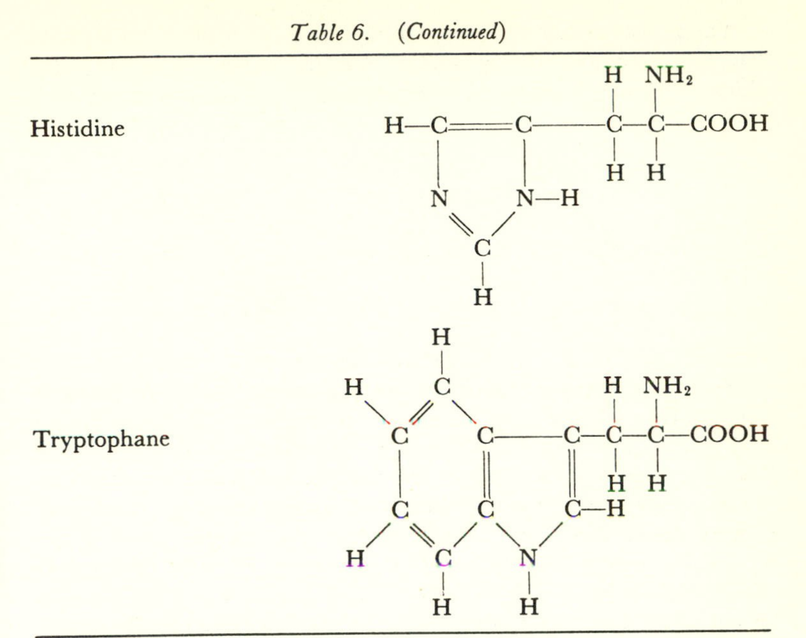

bond needs first attention in considering protein synthesis. If the reaction below is considered to go from right to left, rather than from left to right as is customary, it represents the formation of a peptide. Two amino acids are joined together by splitting off a molecule of H_2O jointly from the carboxyl group of one amino acid and the amino group of the other, to form the peptide linkage. If the reaction is read from left to right it represents the *hydrolysis* of a peptide, that is, the uptake of a molecule of water, resulting in the splitting of the peptide into its component amino acids,

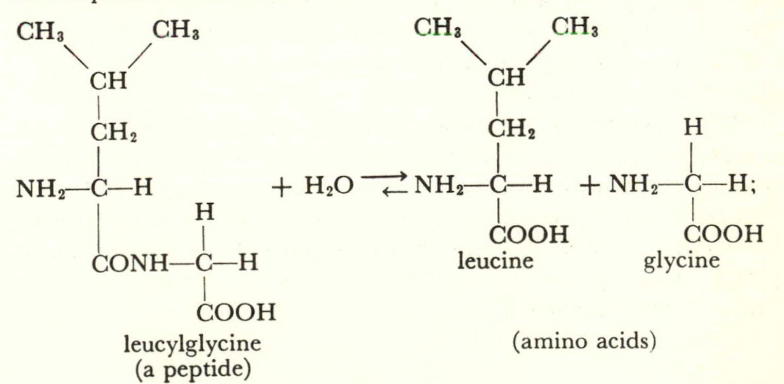

$$\Delta F = -2.96 \text{ kg-cal/mol} \qquad \text{(VIII-1)}$$

The above reaction concerns a dipeptide. Numbers of amino acids may be linked together by peptide linkages to form a polypeptide, which may be represented by the following chainlike arrangement

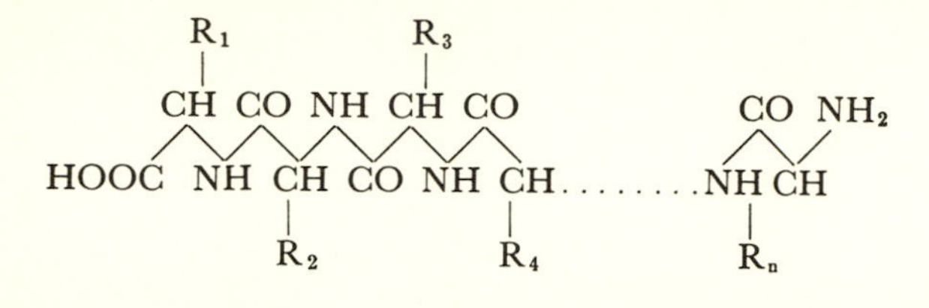

where R_1, R_2, R_3, R_4, R_n, are amino acid groups.

Such long polypeptide chains, held in helical coils by hydrogen bonds, form the "backbones" of protein molecules. The specific amino acid group—that portion which does not take part in the peptide linkage—projects outward from the backbone forming a side chain. A three-dimensional model of the spatial arrangement of the atoms in a short section of a protein backbone, showing the way in which the side chains project above and below, appears as Plate I. In Plate II the model is carried further by including models of the amino acid groups which form these side chains. The side chains provide opportunity for the formation of a variety of *side linkages*, joining together parts of the same or of different backbones, and providing for a variety of protein patterns. Such linkages may involve covalent or ionic bonds, or resonating bonds of intermediate character; hydrogen bonds between oxygen and nitrogen atoms are probably very important. Besides these, Van der Waals forces and attraction at points of opposite electric charge may assist in holding the polypeptide chains together. By determining the positions in which the backbones are held together, the side linkages play a large part in determining the physical and chemical properties of the protein. For example, in the fibrous proteins the polypeptides tend to be straight and parallel, whereas in the globular proteins they tend to be drawn together giving the molecule a more nearly spherical shape.

If the twenty-odd amino acids were combined at random in the polypeptide chains, millions of different kinds of proteins would be possible, but actually there are relatively few general types, indicating that certain restrictions are imposed. The proportion of the various kinds of amino acid group varies widely among the different kinds of protein, the sequence of amino acids in the polypeptide chain following recurring patterns.

Proteins which have been isolated from living systems by methods that presumably leave these compounds unchanged or nearly so are referred to as "native" proteins. These are readily altered by various agents—heat, radiation, chemical agents, etc.—causing them to coagulate. Between the native and coagulated states are intermediate states in which the protein is said to be denatured. Coagulation of proteins is an irreversible process, while denaturation may, under appropriate conditions, be reversed. Thus, we may schematize these steps as follows:

$$\text{native protein} \rightleftarrows \text{denatured protein} \rightarrow \text{coagulated protein}$$

The essential change that takes place in denaturation seems to be the rupture of some of the side chains that hold the polypeptide chains together, allowing them to unfold so that the molecule assumes a spatial configuration different from that of the native protein.

The formation of peptide bonds in building polypeptides presents at least one thermodynamic problem which the living system solves, apparently in a way that would be difficult to follow in the laboratory. The hydrolysis of leucylglycine to leucine and glycine, represented in reaction VIII-1, goes with a decrease in free energy: $\Delta F = -2.96$ kg-cal/mol. The hydrolysis of other peptides involves about the same free energy change, and for purposes of discussion the value -2.75 kilogram-calories per mole will be taken as ΔF for the splitting of the peptide bond. The negative sign of ΔF indicates that the reaction will go spontaneously in the direction of disruption rather than synthesis of this linkage, e.g. from left to right in reaction VIII-1. The equilibrium constant corresponding to $\Delta F = -2.75$ kg-cal/mol is approximately 100, which means that if one starts with pure peptides one may expect to end up with 99 per cent amino acids, only one per cent of the peptides remaining. To form peptides in any quantity from amino acids, then, reactions of the type indicated in VIII-1 must be caused by some means to go in the opposite direction to that which would be taken spontaneously.

The forwarding of endergonic processes by coupling with exergonic ones has been discussed in the last chapter, but this case is of particular interest and may be elaborated a little further. Starting with a mixture of amino acids, about one per cent would be converted spontaneously to dipeptides. This reaction might be accelerated by an enzyme, but the enzyme could not, of course, alter the equilibrium of the reaction. So in the end 99 per cent of the material would still be in the form of amino acids. If, however, the peptide that was formed could be removed by some means, one per cent of the remaining amino acid would be changed into peptide. If the removal of peptide could be accomplished continuously, most of the amino acid would be converted to peptides. This begins to have the appearance of a perpetual motion machine, and indeed it would be one if this cycle were carried out continuously and spontaneously. Actually, however, the continuous removal of peptide requires that work be done upon the system, and some exergonic reaction must supply the energy. It has been shown experimentally that the formation of a peptide linkage directly from amino acids is accomplished by intact cells but not by cells whose structure has been destroyed, and it seems necessary to conclude that the free energy required for the formation of peptide can be provided in some way by the intact living system. While the energy requirement for the synthesis of a single peptide bond is not very great (about 2.75 kg-cal/mol), there may be hundreds or thousands of these linkages in a protein molecule, so the sum total is a factor of some magnitude.

Although not strictly applicable to proteins, the above estimation regarding the free energy change in the formation of polypeptides becomes of interest when we consider the origin of such polymers in a non-living world (see Chapter X). The thermodynamics of the proteins is not quite so simple, since the chains are held together by hydrogen bonds, which should modify energetic relationships. The overall free energy change concerned in the formation of proteins must be positive, however, since they can be hydrolyzed catalytically by appropriate enzymes. The energy of the hydrogen bonds may be thought of as contributing to energy of activation, which lends stability to the protein molecule and must be overcome by catalysis.

Molecules made up of as many atoms as those of proteins could conceivably have an almost infinite number of arrangements of these atoms. Even if one considered that the fundamental unit of the protein was the amino acid, the existence of twenty or more kinds of such units would permit a fabulous number of combinations and permutations, if there were not certain restrictions on their arrangement. A particularly striking restriction in the structure of the amino acids themselves poses a question of fundamental importance. Only α-amino acids, that is, amino acids in which the NH_2 and COOH groups are attached to the same carbon atom, are found in proteins (Table 6), although amino acids with other arrangements may be synthesized *in vitro*. Each α-amino acid with the exception of glycine can exist in two isomeric forms which have spatial arrangements that may be regarded as mirror images of each other

These correspond to the two stereoisomers of lactic acid

one of which, *d*-lactic acid, rotates polarized light to the right, the other, *l*-lactic acid, rotates polarized light to the left. With possible rare exceptions, the amino acids that are found in proteins are all of the *l* form, although, since the optical rotation depends upon the structure of the molecule as a whole, some of them rotate polarized light to the right. If an optically active substance is synthesized in the laboratory a mixture of *d* and *l* forms is obtained, such a mixture being the most probable because there is thermodynamically no difference between the *l* and the *d* forms, the free energy of formation being the same for both. Separation of the two optical isomers from a racemic mixture requires that work be done, as in the famous manual separation of *d* and *l* tartaric acid by Louis Pasteur. Thus, whereas one would expect a spontaneous reaction to yield both in equal quantities, the living organism builds only one kind of isomer. The ability to form only one of two stereoisomers is found not only in the formation of amino acids, but elsewhere in living systems. The basic stereoisomerism of the amino acids is associated with the same carbon atom that forms the peptide linkage, and it may be that the long polypeptide "backbones" of the proteins can only be built up from amino acids having the same stereo-configuration. There seems no obvious reason, however, why some of the proteins should not be made up from *d*-amino acids instead of all being "left-handed"—unless this reason is an evolutionary one. The restriction to a single stereoisomeric type suggests, at any rate, that the proteins have to be built up according to some spatial pattern which accommodates only one of the two mirror image forms.

Nucleic Acids

Nucleic acids play their great role in living systems as means of coding and transfer of information. This includes providing patterns of protein synthesis. These long polymers are of two general types: *deoxyribose nucleic acid*, DNA, and *ribose nucleic acid*, RNA. Information is carried from one generation of cells to the next by DNA, which may in a sense be thought of as furnishing the master plan of inheritance. RNA is more directly concerned in the synthesis of proteins and in carrying information from one part of the cell to another.

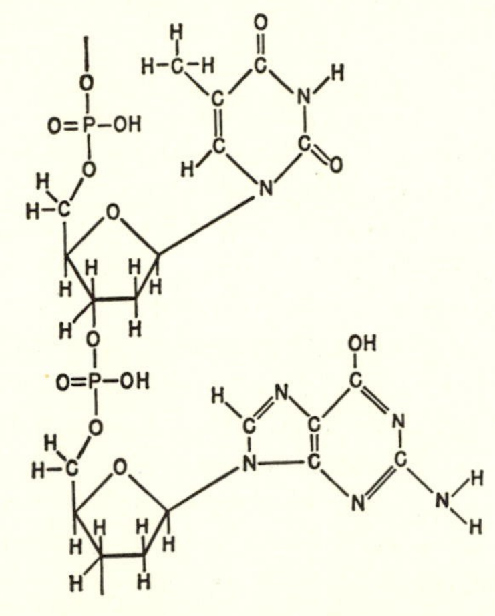

FIGURE 16. Two deoxyribose nucleotides joined as in DNA. Above, thymidine nucleotide, which has a pyrimidine base; below guanidine nucleotide, which has a purine base. In the joining of the two nucleotides, one molecule of water is removed.

The nucleic acids are long polymers composed of nucleotides joined together in a chainlike fashion which recalls the proteins. The joining together of two nucleotides is shown in Figure 16. Each nucleotide is composed of a purine or pyrimidine base, (Table 11), a five carbon sugar, and a phosphate group. Adenylic acid, ADP, and ATP, which we met in the last chapter, are nucleotides. Like adenylic acid the nucleotides in nucleic acid have only one phosphate group. DNA and RNA are distinguished by the kind of sugar in the nucleotides composing them—deoxyribose in the former, ribose in the latter; the nucleotides shown in Figure 16 are of the deoxyribose type, ATP is of the ribose type.

In DNA the polymer chains are coiled in a double helix, being held in this form by hydrogen bonds between purine and pyrimidine bases; we owe the discovery of this double helix structure to J. D. Watson and F. H. C. Crick. Only four types of purine and pyrimidine bases are found in DNA; adenine, cytosine, guanine and thymine. The arrangement of nucleotides with these bases in the helical structure, according to strict rules, provides a code in terms of which information for replication may be recorded. RNA has the pyrimidine base, uracil, instead of thymine; the coiling of the nucleotide chains is less well understood than in DNA.

As in the case of the proteins, the joining together of the monomers to form nucleic acids involves the removal of water. As this reaction takes place in the aqueous medium of the cell it would be expected to entail a high positive free energy, since many molecules of water must be removed in building up one of these long chains. The nucleic acids can be hydrolyzed catalytically into their monomers by appropriate enzymes, indicating that this process is exergonic, and that the reverse reaction must involve the expenditure of free energy somewhere in the system. Polynucleotides may be formed *in vitro* from di- and triphosphate nucleotides if appropriate enzymes are present, and here the free energy requirements may be satisfied by the loss of energy-rich phosphate. It may be concluded that the building up of long, nucleic acid polymers *in vivo* is, like the building up of proteins, dependent upon the free energy supplied in one way or another by cell metabolism.

Replication

When a living cell divides the result is, as a rule, two quite exact copies of the parent cell; and when these two daughter cells in turn divide they produce copies of themselves and of their parents. The copying is not always strictly exact—chance mutation is of greatest importance in evolution, and in the development of multicellular organisms the cell characteristics change in orderly manner, a process called differentiation. But although striking and important in their results, such modifications may be thought of as minor alterations, and for the immediate discussion we may assume that replication of living systems is quite exact. The information for the replication of the complex cell system is generally thought of today as being coded in DNA in the cell nucleus. On the basis of this information, mediated by RNA, the various proteins are put together from the correct amino acids, and these then play their roles as enzymes in the cell metabolism. We may picture here a complimentarity between nucleic acids and proteins—close associations of the two being often designated as nucleoproteins. In an over-all view, perhaps other kinds of molecules contribute a modicum of information essential to replication. In any case, the replication of any known kind of cell involves a complicated arrangement of parts at both intra- and inter- molecular levels that is beyond our present powers of describing or copying.

To help in thinking about the mechanism of replication in physical terms, analogy has often been had to a template, such as is used in mechanical operations to reproduce spatial dimensions and relationships. This has been a fruitful analogy which may still be useful; but it tends to emphasize spatial aspects too much, perhaps leading us to picture molecules such as the long polymers of protein and nucleic acid as being built up directly against previously existing molecules. With our increasing knowledge of the means of coding and transfer of information for replication, such ideas should not be taken too literally.

Replication of living systems may be thought of as having at least three aspects: a spatial aspect which has to do with arrangements of molecules or molecular aggregates with respect to each other; an energetic aspect having to do with assembling these component parts and holding them in their spatial relationships; and a kinetic aspect having to do with the rates of processes concerned in putting the parts together. Clearly the three aspects cannot be separated within the living system, although one may learn something about them by isolating various parts, for example, the kinetic function of the proteins. For we cannot break living systems into their component parts and still have living systems. The latter not only contain all the component molecules, but are capable of maintaining these in their proper relationships and of assembling new components for the replication of the whole. Some things that are often thought of as living do not fall within this definition, for example, neither mammalian spermatozoa nor viruses may be so classed, although the latter are often referred to popularly as "living molecules"; neither of these systems—the one cellular the other not—is capable of mobilizing the necessary free energy, although each contains replicable information, coded in its nucleic acids. In both these cases replication involves the metabolism of a complete living system—the egg and the "host" cell, respectively—to fulfill energetic and kinetic requirements.

Such thoughts enter when one tries to picture the emergence from a non-living world of systems capable of metabolism, replication, and organic evolution. These are matters that will be taken up in Chapter X and again in Chapter XIII.

IX · STABILITY AND VARIABILITY

"... *natural selection can only act on the variations available, and these are not, as Darwin thought, in every direction.*"—J. B. S. HALDANE

IF one wishes to push his criteria of likeness far enough, he may say that no two living organisms are ever exactly alike. Probably no offspring is ever just like its parent, although in general the resemblance to parents is greater than to the other members of the species; and the resemblance between members of a species is greater than that between the members of two distinct species. Thus, among living organisms, while difference is the rule, the differences are circumscribed within the limits of certain groupings, and this makes possible the separation of one species from another. At another level, species having certain characteristics in common may be placed together in larger groups, and these assembled into still more inclusive groups, to form a logical system of classification. There may be room for quibble as to the exact lines to be drawn between species, and as to other details of arrangement; but there can be no question that an orderly relationship exists among all forms of life. Such relationship finds its explanation in common ancestorship and evolution, and modern classifications are universally based on this concept.

The persistence of well-defined species indicates a certain essential degree of stability of biological pattern. That is, while there is constant change in the organism, with temporary modification, destruction, and renewal of parts, the same overall pattern is maintained in great detail, and is passed on from parent to offspring. Yet evolution could occur only if there were a certain amount of variability in this pattern; otherwise species would remain permanent and immutable. Moreover, for evolution—in the sense of modern Darwinism at least—there must be some variants from the general pattern that are themselves relatively stable; that is, once such variants occur, they must be handed on to successive generations without reverting too readily to the ancestral type. Such persistent variant patterns are essentially what we distinguish as *mutations;* and these, according to modern theory, provide the basis for natural selection. That is, a mutant form if better adapted to the environment has better chance of survival, and is hence "selected." It seems obvious that too radical departure from the ancestral pattern would not be compatible with life; and that unstable variants—persisting for, say, a single generation and then reverting to the ancestral pattern—would not have evolutionary significance. Such essential stability with minor variability is embodied in the mechanism of heredity of modern organisms.

The Mechanism of Heredity in Modern Organisms

The gene theory of heredity is intimately associated with the cell nucleus. In the classic form of that theory, the genes, which are the determiners of heredity, are pictured as essentially particulate and as located at definite positions in the chromosomes. In the cells of higher plants and animals the chromosomes are elongate structures identifiable in certain of the pictures shown in Plates III and IV, which illustrate stages in the process of mitosis, the most common type of cell division. The chromosomes are present in pairs, each member of a pair having the same size and shape, with the exception of an unequal pair, the sex chromosomes, one of which may be absent entirely. At a certain phase of mitosis the chromosomes line up side by side as shown in the figures. Each chromosome splits lengthwise, and when the cell separates into two at the end of mitosis, each of the daughter cells carries with it a full complement of chromosomes. At times certain cells go through a series of divisions in such a way that cells are produced which have only one of each pair of chromosomes, the haploid number. These are the *gametes* or sex cells. The process by which they are formed, called *meiosis*, is not illustrated in the pictures. Two gametes may fuse to form a *zygote*, or fertilized egg, which develops into a new organism. As a result the zygote contains all the chromosomes of the two sex cells, the diploid number. All the cells deriving from the fertilized egg by mitosis—which make up the major part of the multicellular organism—also contain the diploid number of chromosomes. According to the gene theory each chromosome contains many genes, each located at a certain position in the chromosome; and each pair of chromosomes, with the exception of the sex chromosomes, contains the same number and kind of genes, i.e. pairs of genes determining the same inherited characters are shared between the paired chromosomes. Commonly one gene of a pair dominates in determining the particular character displayed by the individual, so we speak of *dominant* and *recessive* genes. In the formation of the

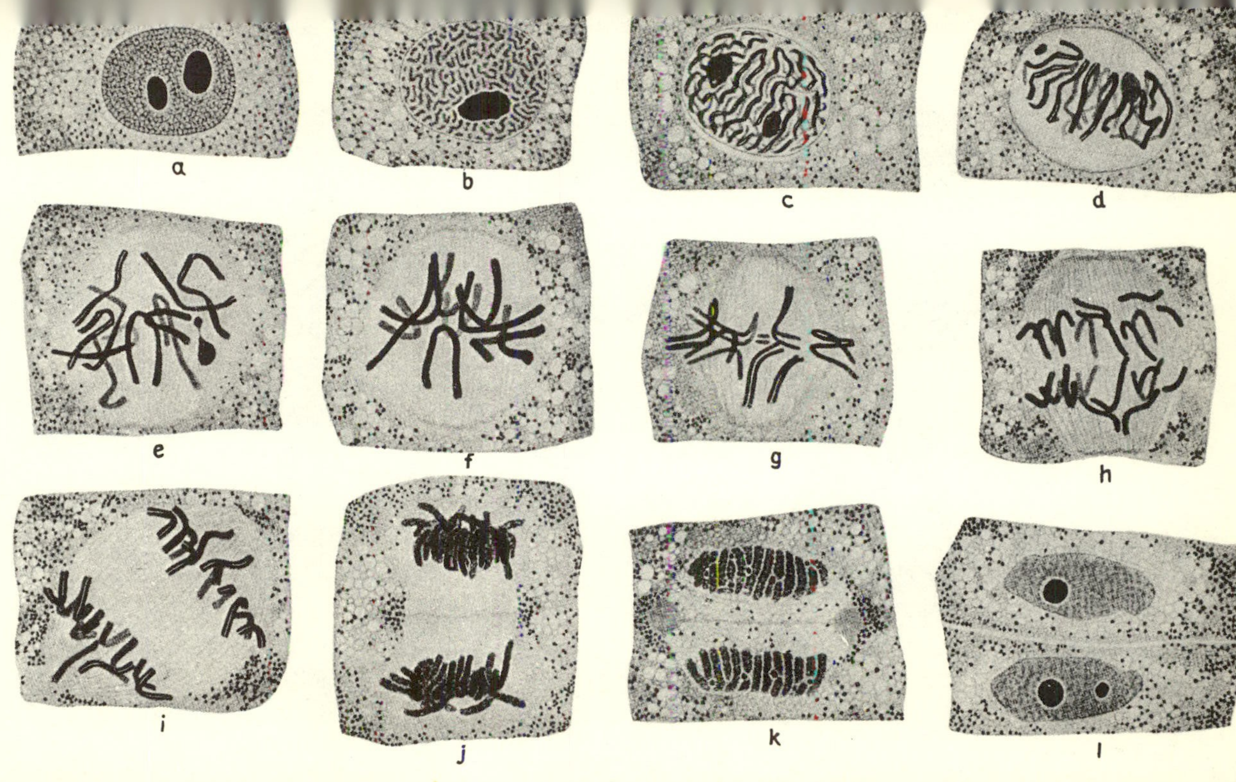

PLATE III. Stages in cell and nuclear division in the root tip of the onion (*Allium cepa*). In the resting stage (a) no chromosomes are recognizable; they are first suggested in prophase (b-c). In metaphase (f) the chromosomes are clearly distinguishable, and in anaphase (g-i) they pair up, split and move apart. In telophase (j-k) the two sets of daughter chromosomes begin to lose their discrete character, and in the two daughter cells (l) are no longer distinguishable. From Bělăr, K. Der Formwechsel der Protistenkerne. *Ergebnisse und Fortschritte der Zoologie* (1926) *6*, 235-648.

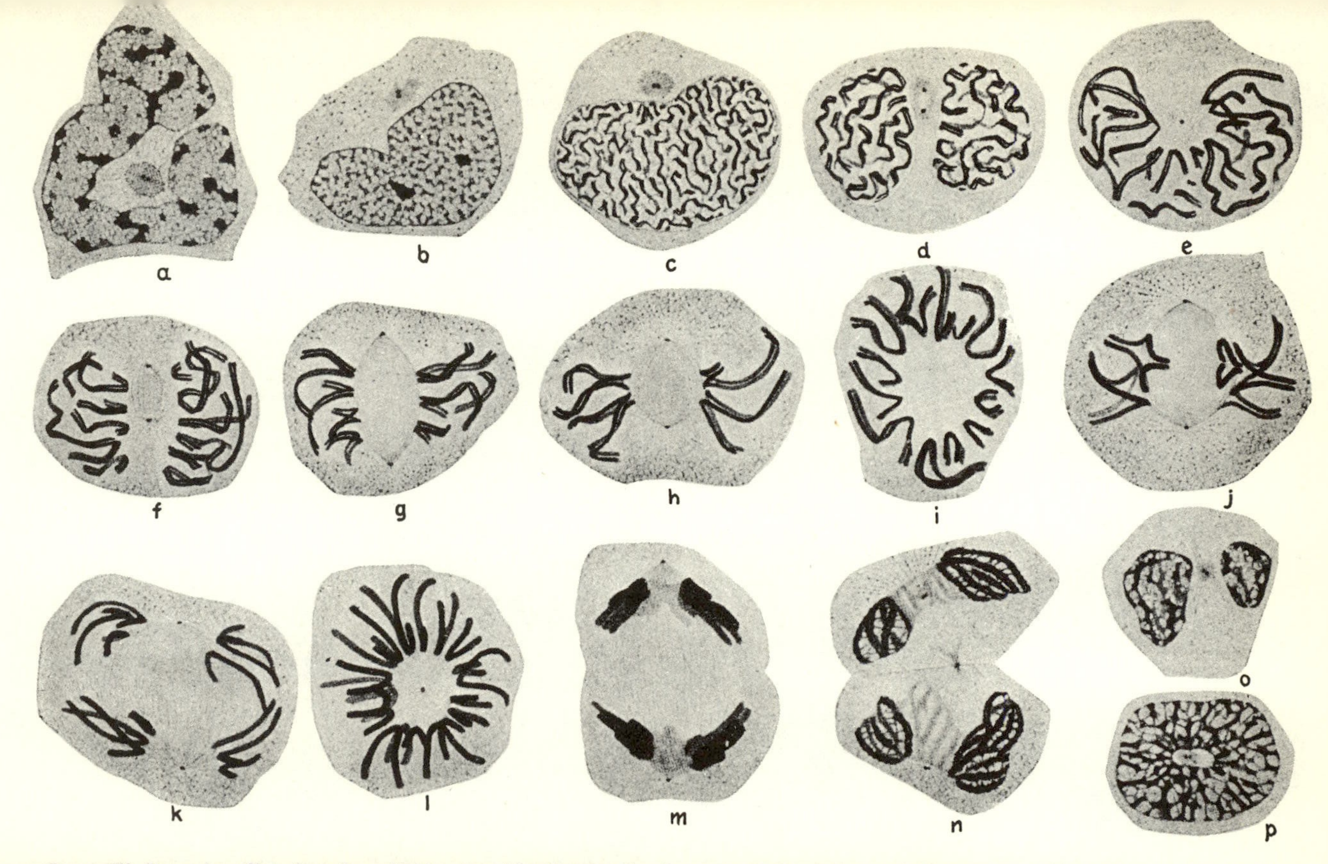

PLATE IV. Stages in cell and nuclear division in white blood cells of a newt (*Salamandra maculosa*). Stages comparable to those in the

gametes in meiosis, each gamete receives only one of each pair of chromosomes, and hence one gene for each hereditary character, whereas the zygote has two. The chromosomes are shuffled in meiosis, and as a result all gametes do not carry the same genes; the hereditary equipment of the zygotes is thus subject to chance—acting within certain limits—since those genes located in the same chromosome are usually passed on together in the gametes. The possibilities of a gene crossing over from one chromosome to another, of rearrangement of

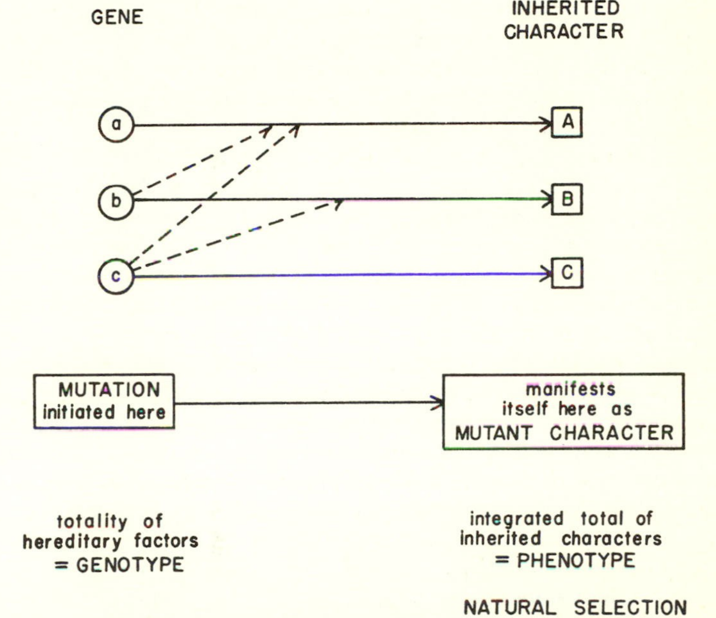

FIGURE 17

parts of the chromosomes thus effecting new spatial distribution among the genes of the offspring, and of mutation of the genes add to the complexity of this gambling device bringing other elements of chance into the hereditary make-up of the individual.

The distinction between gene and inherited character needs to be clearly drawn at this point. Figure 17 illustrates this relationship schematically, in a general way. On the left-hand side of the diagram three genes *a*, *b*, and *c* are represented by circles; while their overt expressions, the inherited characters, are represented on the right by the squares *A*, *B*, and *C*. The arrows connecting gene and inherited character are intended to represent series of reaction steps intervening between the two. The direction of the arrows indicates that whereas the gene determines the nature of the inherited character, the character has no direct influence in determining the nature of the gene. Each principal arrow leads from a gene directly to an inherited character, this being intended to indicate that the particular gene has a predominant effect upon the particular character. Less predominant influence of certain genes is indicated by broken arrows. The multiple influence of genes on a given character probably represents the more common case. The totality of the genes is referred to as the *genotype;* the integrated total of the inherited characters as the *phenotype.* Different genotypes may find expression in the same phenotype. For example, the inherited character may be the same if one of a pair of genes is dominant and the other recessive, as if both are dominant.

A mutation may be thought of as an actual change in the gene itself, occurring only rarely. But the occurrence of the mutation is only detected by some change in an inherited character, and this will be referred to as a *mutant character*, to distinguish it from the underlying change in the gene itself which is the *mutation.* In most cases the mutation manifests itself in the modification of some observable aspect of the organism rather than in a complete loss of a measurable character, although the latter may occur. The scheme in Figure 17 indicates that the obliteration of a reaction pattern by the mutation of the gene *c* might result in the loss of characters *A* and *B*. Perhaps the majority of mutations are lethal, that is, they result in the death of the individual, whether early or late in its development. Breaks in chromosomes resulting in rearrangements, duplications, deletions of parts, and so forth are sometimes referred to as mutations, but perhaps more commonly as *chromosome aberrations.* The interchange of parts of chromosomes during the process of meiosis is known as *crossing over.* This is a more common phenomenon than mutation or chromosome aberration. In crossing over a number of genes may be transferred together, and this has provided an important tool for determining the positions of genes in the chromosome, particularly in the fruit fly *Drosophila*, for which accurate chromosome "maps" are available.

The biochemical approach to genetics has received great impetus recently, particularly in studies on microorganisms. The method introduced by G. W. Beadle and E. L. Tatum in their studies on the bread mold *Neurospora* has contributed a particularly basic approach, and a typical procedure will be briefly described. Spores of the mold are treated with mutagenic agents, ultraviolet radiation for example, in doses sufficient to kill a considerable proportion of the organisms. The survivors are then grown upon a "complete" medium which contains a wide variety of organic substances the fungus may require, and various strains are isolated. The wild type *Neurospora* can grow on a minimal medium containing only sugar, inorganic nitrogen, salts, and the sulfur-containing compound biotin; from these simple substances the organism can synthesize whatever compounds it needs for maintenance and growth. Some of the strains resulting from treatment with mutagenic agents grow on the complete medium, but fail to grow on the minimal medium, having presumably lost by mutation the ability to synthesize some of the substances not provided in the minimal medium. Next the attempt is made to discover the nature of the lost synthetic ability by supplying critical substances one at a time. If the minimal medium plus a specific substance, say an amino acid, supports the life of a given mutant strain, it may be assumed that mutation has resulted in the loss of a key step in the process by which that particular substance is synthesized. By application of this technique a considerable number of strains of *Neurospora* have been isolated, each lacking ability to synthesize some specific amino acid or other substance. The method has also been successfully adapted to the study of other microorganisms by various workers. Such experiments indicate a rather simple and direct relationship between gene and inherited character (as suggested by the solid lines in Figure 17), and the potentialities for the study of the mechanism of heredity presented by such methods can hardly be overemphasized. The relationship seems best explained by the assumption that the gene provides the essential information for replicating an enzyme necessary for the synthesis of the particular amino acid or other substance. If a mutation of this gene changes the pattern and prevents the reproduction of the necessary enzyme, the ability to carry out this specific synthesis is lost.

Such biochemical studies indicate the one-to-one nature of the relationship between genes and chemical reactions. In the synthesis of a complex substance a number of genes may be concerned, each determining a given step. In the synthesis of the amino acid, arginine, by *Neurospora*, at least seven genes are concerned, the inactivation by mutation of any one preventing the synthesis of this substance.[1] This indicates the existence of at least seven reaction steps in the synthesis. While arginine is an important constituent of proteins, two of the intermediate substances in its synthesis were found to be citrulline and ornithine, amino acids that do not appear as constituents of proteins, and are not known to be used directly in the economy of the organism.

[1] Srb, A., and Horowitz, N. H., *J. Biol. Chem.* (1944) *154*, 129–139, Horowitz, N. H., Bonner, D., Mitchell, H. K., Tatum, E. L., and Beadle, G. W., *Am. Naturalist* (1945) *79*, 304–317.

From such studies it appears that the arrows in Figure 17 should be thought of as composed of series of reaction steps, each step involving a single gene. The interplay of many genes in determining a single hereditary character is thus more readily explainable.

Stability and Variability

When the concept of the gene was originally formulated at the beginning of this century it was no doubt thought of as being more or less particulate. This idea still serves conveniently for many purposes. We know now that genetic information is carried in the DNA component of the chromosomes, and we may picture the gene as a locus in this material where a specific message is coded. The long, chain-like character of the DNA molecules accords with the need for both general stability and the possibility of minor variation of genetic information that are essential if evolution is to occur. We may think of a mutation as a change in the information coded in a portion of such a chain—say the rearrangement of a small number of monomers—which takes place without disruption of the whole polymer. The stability seems to extend also to relationships between polymer molecules, as indicated by the exchanges of sections of chromosomes carrying many genes (chromosome abberations) with only minor change in the whole of the genetic information.

A Thermodynamic and Kinetic Model of Mutation

It may be instructive to picture mutations as chemical reactions and to treat them according to the rules of thermodynamics and kinetics, although quite arbitrary assumptions must be made in order to do so and the available data may not be strictly applicable. Let us think of mutation as an isothermal chemical reaction in which this molecule undergoes minor alteration in properties not affecting its general character, let us say, the rearrangement of a single bond at some specific place in the molecule. We may represent this reaction as follows

$$\mathrm{N} \rightleftarrows \mathrm{M} \qquad \text{(IX-1)}$$

where N is a "normal" molecule and M is a "mutant" molecule. Equilibrium between the forward and reverse mutation is then defined by the ratio of the forward, k_f, and reverse, k_r, rates

$$K = \frac{k_f}{k_r} \qquad \text{(IX-2)}$$

K being the equilibrium constant. Knowing the value of K the free energy change (ΔF) of mutation could be calculated from equation

(III-4). These functions deal with equilibria, but give no information about the rate of achieving equilibrium, or how often a mutation may be expected to occur. The rate of chemical change is a function of another quantity, the free energy of activation, $\Delta F^{\ddagger}$, which could also be calculated in the present instance if appropriate information regarding the rate of mutation and its change with temperature were available. The relationship between these functions is illustrated diagrammatically in Figure 3, where ΔF would represent in the present instance the difference in free energy between the normal and mutant states, and $\Delta F^{\ddagger}$ the difference in free energy between the normal and an activated state which must be reached before the normal molecule can become a mutant molecule.

If we may carry this analogy over to the gene, the same kind of relationships should apply. We can imagine the gene mutation to be

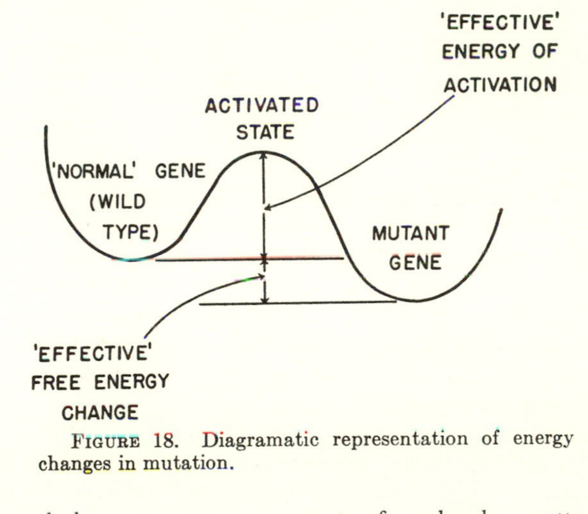

FIGURE 18. Diagramatic representation of energy changes in mutation.

represented by some rearrangement of molecular pattern. For various reasons, including difficulties in calculations such as will be discussed shortly, it seems best to have a somewhat less restricting diagram than Figure 3 to represent the act of mutation, and one is suggested for this purpose in Figure 18. Here the terms *effective free energy* and *effective energy of activation* replace the more exact ones in Figure 3, but otherwise the diagrams are comparable. One sees normal and mutant genes as occupying different energy levels, separated by an energy barrier which must be surmounted before transition from one to the other is achieved.

With this picture in mind we may attempt to estimate thermodynamic and kinetic properties of mutation, using data that are available from experiments. It is unlikely that any definite idea of the mechanism of mutation is to be gained from these calculations, but it may be possible as a result to view certain aspects of the process more clearly. The data we shall use are not strictly adequate for the quantitative estimates that are attempted; but they will serve to illustrate certain points. They will at least serve to indicate the difficulties that would be encountered in obtaining data really suitable for such treatment. One of the most serious difficulties is that one cannot measure the incidence of mutations directly, but only in terms of mutant characters. The complexity of the relationships involved is suggested in the diagram in Figure 17, where the interplay of genes in determining a given hereditary character is indicated; the scheme, of course, minimizes the true complexity of the relationships.

Our model reaction, involving the rearrangement of a single bond in a molecule, is a monomolecular one. We shall assume that gene mutation is a reaction of the same type, this being about the only feasible way to treat mutation as a chemical reaction. This figment receives some support from experiments on the induction of mutations by radiation. Certain quantitative relationships are involved which may need to be made apparent. A mutation is thought of as involving change in only a single molecule in a gamete. This mutation is only manifest as an alteration in a characteristic of the organism which develops from the zygote to which this mutant gamete contributes. In a given number of adult organisms which we may choose to examine, this mutant character will be observed in only a certain proportion. On the basis of our hypothesis we assume that the number of molecules that have undergone the specific chemical change corresponding to this mutation is the same as the number of organisms in which we observe the mutant character. This is certainly an indirect way of obtaining data for thermodynamic calculations, but seems sound if the basic postulates are accepted. In the following treatment the total number of gametes will be indicated by the symbol g, the number of mutants by m, and these quantities will be treated as though they represented numbers of molecules. Thermodynamic quantities will be expressed, as elsewhere, in kilogram-calories per mole.

On the basis of these assumptions the rate of mutation may be defined by the equation

$$k = \frac{m}{gt} \qquad \text{(IX-3)}$$

where g is the total number of gametes concerned, m the number of mutants occurring in time t, and k is the rate constant. Substituting from (IX-3) in (IX-2), using subscripts f and r for forward and reverse as before,

$$K = \frac{m_f g_r t_r}{m_r g_f t_f} \qquad \text{(IX-4)}$$

For a given period of time, represented in the instance to be cited by the breeding span of the animal, $t_f = t_r$, and

$$K = \frac{m_f g_r}{m_r g_f} \qquad \text{(IX-5)}$$

This equation may now be applied to a specific case. The only data that seem suitable for the purpose are those of N. W. Timoféeff-Ressovsky who measured the frequencies of forward and reverse mutations of a few genes having to do with eye color in the fruit fly *Drosophila*. Values of K calculated for these data from equation IX-5 appear in Table 7. In these experiments the mutations were induced by treat-

Table 7. Equilibrium of Mutations in *Drosophila melanogaster*[1] (larvae treated with x-ray, 4,800 to 6,000 r units).

Mutation[2]	Forward		Reverse		K	ΔF
	No. of gametes	No. of mutations	No. of gametes	No. of mutations		kg-cal/mol
$f^+ \rightleftarrows f$.	43,000	11	44,000	15	0.75	+0.17
$p^+ \rightleftarrows p$	52,000	1	58,000	9	0.12	+1.26
$w^+ \rightleftarrows w^e$	69,500	9	72,000	3	3.11	−0.67
$w^e \rightleftarrows w$	69,500	28	54,000	0	>21.75	<−1.82
$B^+ \rightleftarrows B$	69,500	0	9,000	8	< 0.016	>+2.45

[1] Experimental data from Timoféeff-Ressovsky N. W., *Ztschr. Induct. Abstamm- und Vererblehre* (1933) *64*, 173–175, *65*, 278–292; also in *Nachr. Ges. Wiss. Göttingen* (1935) N.F. Fachgruppe VI, p. 213; and *Experimentelle Mutationsforshung in der Vererbungslehre*, Dresden (1937) Steinkopf.

[2] f = forked bristle, p = pink eyed, w = white eyed, w^e = eosin eyed, B = bar eyed. The superscript + indicates the wild type in each case. For an introduction to nomenclature and symbols see Sturtevant, A. H., and Beadle, G. W., *An Introduction to Genetics*, Philadelphia (1939), Saunders.

ment with x-rays, in order to obtain them at measurable rates, and hence, to relate the values of K to spontaneous mutation the assumption must be made that the forward and reverse rates are accelerated to the same relative degree by this treatment. This is an assumption commonly made, but to which serious objection might be raised. Indeed, there is reason to think that the mutations studied by Timoféeff-Ressovsky may not represent strictly reversible changes, as assumed in the following treatment; and all in all it would be unwise to place too much faith in the actual quantities calculated. Nevertheless, these data serve for purposes of illustration, and the uncertainty as to their genetic interpretation need not vitiate our discussion of general principles so long as we are aware that uncertainties exist.

Treating K as the equilibrium constant for a chemical reaction, the free energy change, ΔF, may be calculated according to equation III-4. Values so obtained are presented in Table 7. The first three items all represent very small free energy changes. In the case of the fourth no reverse mutations were actually observed, but are assumed to occur for calculation of an apparent limiting value for the free energy change. In the fifth item no forward mutations were observed, but the apparent limiting value has again been calculated. Since, in a thermodynamic sense, the direction of the reaction is arbitrarily taken (I have followed the directions assigned by the author), the sign of ΔF may be disregarded and only its magnitude considered.

The calculations indicate that for these particular mutations the free energy changes are not very great, and hence that these processes may be regarded, from the thermodynamic point of view, as readily reversible. In fact this could almost be said for any mutation which was known to go in both directions, since it may be estimated that mutations characterized by values of ΔF much higher than those indicated in Table 7 would be so infrequent that the chance of observing them in experiments involving numbers of animals feasible to study would be very small. Table 8 will assist in illustrating this point. The free

Table 8.

Chemical Reaction	ΔF kg-cal/mol (approx.)	Number of mutations in the direction of lower free energy, per mutation in the opposite direction, corresponding to this ΔF.
Hydrolysis of peptide bond	−2.75	1×10^2
ATP → ADP	−10	2×10^7
Combustion of $\{CH_2O\}$	−120	5×10^{87}

energy changes involved in three reactions with which we have become acquainted in earlier chapters are shown therein, and corresponding to each of these the relative numbers of mutations that should occur in opposing directions if the same free energy were involved. It is seen that a mutation involving 2.75 kilogram-calories per mole—corresponding roughly to that for the hydrolysis of the peptide linkage, and the highest ΔF calculated in Table 7—would occur only once in one direction to 100 times in the opposite. Assuming that the more frequent mutation occurred once in 1,000 gametes, the opposite mutation would occur only once in 100,000. The assumed proportions of mutations correspond to some of those shown in Table 7, but it must be remembered that in those experiments the mutations were speeded up by x-rays, and in nature the rate of mutation is considerably lower. Obviously the chances of demonstrating reversal of a mutation involving a free energy change much greater than this one would be small indeed. If the free energy change of a mutation corresponded to that for the loss of a phosphate bond by ATP—about four times that for hydrolysis of the peptide linkage—mutation in the least frequent direction would occur only once to over one-hundred million in the opposite direction. If a mutation involved a free energy change corresponding to that for the oxidation of a unit of carbohydrate, $\{CH_2O\}$, the possibility of mutation in one direction would be too small even to consider. If we think of mutation of, say, a minor rearrangement of a DNA molecule, we realize too that there are probably not many reactions involving high free energy changes that could take place without bringing about more or less complete disruption of the molecule. Conceivably, reactions involving free energies as high as 10 kilogram-calories per mole might take place, however, without such disruption. But even the latter could give a picture of apparent irreversibility.

The above treatment might lose significance, although the thermodynamic problem would remain, if it could be proved that—as frequently asserted at present—reversal of mutation in the sense we have postulated does not occur. The question rises largely from the difficulty of knowing whether the return of a mutant character of the phenotype to its normal counterpart necessarily represents a direct return to exactly the same genotype. That is, does the outwardly manifest reversal represent the direct reversal of an alteration in a single gene? Calculations of the type shown above suggest that this may indeed be a difficult question to answer, for if true reversal occurred it might in many cases be very difficult to demonstrate, because of the tremendous number of observations required.

It has already been pointed out that if mutation is to be regarded as a chemical reaction, its rate should bear no direct relation to its incidence or its reversibility. Let us examine the rate problem briefly. The equation for specific reaction rates given as III-19 is rewritten here as it might apply (to a close approximation) to a monomolecular reaction such as has been postulated

$$k \cong \frac{\mathbf{k}T}{h} e^{-\Delta F^{\ddagger}/RT} \qquad \text{(IX-6)}$$

This is similar in form to the more familiar "Arrhenius" equation

$$k = Ae^{-E/RT} \qquad \text{(IX-7)}$$

from which modern theory of reaction rates originally derives. Here E is the experimental energy of activation, and A is a constant. The experimental energy of activation (sometimes symbolized as μ) is often taken as characterizing a particular biological process. There is fallacy in this practice since E may differ widely from the free energy of activation $\Delta F^{\ddagger}$, approaching more closely $\Delta H^{\ddagger}$. The relationship between $\Delta H^{\ddagger}$ and $\Delta F^{\ddagger}$ is a fundamental thermodynamic one corresponding to III-8,

$$\Delta F^{\ddagger} = \Delta H^{\ddagger} - T\Delta S^{\ddagger} \qquad \text{(IX-8)}$$

where $\Delta S^{\ddagger}$ is the entropy of activation. Within the range of temperatures at which biological processes ordinarily take place, E may be close to $\Delta H^{\ddagger}$, in which case we may write the approximate relationship,

$$\Delta F^{\ddagger} \approx E - T\Delta S^{\ddagger} \qquad \text{(IX-9)}$$

From (IX-6) and (IX-9),

$$k \approx \frac{\mathbf{k}T}{h} e^{-E/RT} e^{\Delta S^{\ddagger}/R} \qquad \text{(IX-10)}$$

If the logarithm of k is plotted against the reciprocal of the absolute temperature (the common "Arrhenius plot") a straight line is usually obtained, the slope of which gives E. This has been done in Figure 19 for data from certain experiments on *Drosophila* to show that reasonably satisfactory values of E can be obtained. In interpreting these data it has been assumed that mutations may take place at any time from the hatching of the fly to sexual maturity, and this "development" time has been used as the basis of determining k from equation IX-10. It is clear from observations by Timoféeff-Ressovsky that the rate of mutation is not constant throughout this time, but this objection may be met by saying that we deal with an average rate over that period, and are permitted a reasonable approximation. The development time varies with the temperature, but correction for this can be made on the basis of data on *D. melanogaster* obtained by Timoféeff-Ressovsky and by Harold H. Plough.[2] The data on unstable miniature in *D.*

[2] *Biological Symposia* (1942) *6*, 7–20.

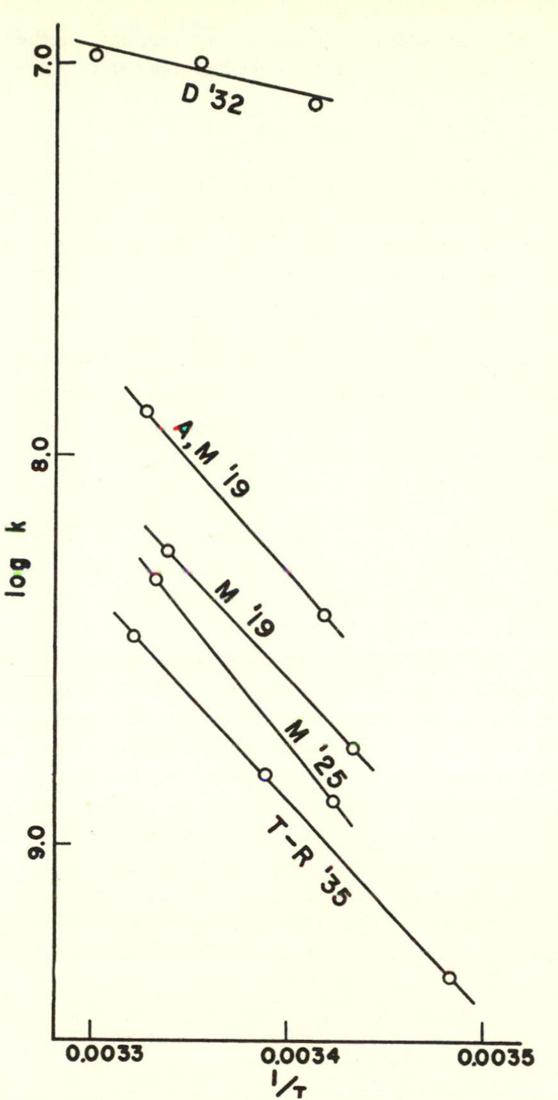

FIGURE 19. Temperature characteristics of mutations in the fruit fly (*Drosophila*).

Drosophila melanogaster

Lethals in Chromosome I

A,M '19—Altenburg and Muller 1919 (Muller, H. J., *Genetics* (1928) **13**, 279–357).

T-R '35—Timofeeff-Ressovsky 1935 (Timoféeff-Ressovsky, N. W., *Experimentelle Mutationsforschung in der Vererbungslehre.* Dresden, (1937) Steinkopf.)

virilis have been corrected in the same way although another species is involved; the correction yields a somewhat more satisfactory plot than is obtained from the raw data.

From the slopes of the curves in Figure 19 values of E may be obtained, and from values of k calculated from the experiments we might substitute in IX-10 and IX-9 to obtain values for $\Delta S^{\ddagger}$ and $\Delta F^{\ddagger}$. But would these values have any real meaning? Granted the validity of the assumptions made to obtain IX-10, which may be open to question, there is serious objection to using the data for lethal mutations to obtain values of k, because the method by which these mutations were measured gives a combined rate for all lethals in a given chromosome. The rates for single lethal mutations should be considerably lower, and this would of course affect the calculated values of $\Delta F^{\ddagger}$ and $\Delta S^{\ddagger}$. In fact, one of the things such calculations illustrate most clearly is the inappropriateness of quantitative treatment of such data as are available, and the difficulty of obtaining any that would be rigorously interpretable in this way.

There is also a theoretical objection that may be raised against calculations of this kind regarding the kinetics of mutation. All the above calculations involve the assumption that the gene is an isolated molecule, whereas it is really part of a reaction system. In the living organism, protein molecules continually undergo change, constituent atoms being replaced. Into the maintaining of the stable pattern of a polymer molecule in spite of this constant change may go a certain amount of free energy derived from the metabolism of the organism. This would introduce a factor not represented in equation IX-6. It might be expected that the amount of rebuilding, and hence the amount of energy introduced would vary widely among organisms whose life spans are very different, e.g. *Drosophila* and Man, and in such case calculated values of energies of activation would have little quantitative meaning. This objection and some of the others mentioned may apply to the analyses of others who have made somewhat different approaches to the problem.[3]

[3] For example Timoféeff-Ressovsky, N. W., Zimmer, K. G., and Delbrück, M., *Nachr. Ges. Wiss. Göttingen* (1935) N.F. Fachgruppe VI, 189–245; Schrödinger, E. *What is Life*, New York (1945) Macmillan.

Lethals in Chromosome II
M '19—Muller 1919–20 (Muller, H. J., *Genetics* (1928) **13**, 279–357)
M '25—Muller 1925–26 (Muller, H. J., *Genetics* (1928) **13**, 279–357)
Drosophila virilis
Unstable miniature (somatic)
D '32—Demerec 1932 (Demerec, M., *Proc. Nat. Acad. Sci.* (U.S.) (1932) **18**, 430–434)

The above calculations have involved numerous assumptions which may not be valid, and one might arrive at the model presented in Figure 18 without so elaborate a treatment. If mutation is to be considered as a chemical reaction, or analogous to one, a general scheme of this kind should describe the energetic relationships, and—in spite of the difficulties which may be met in attempting a quantitative evaluation—we should not be led too far astray by reasoning in such terms. Proceeding a little further with this analogy, it might be predicted that the rate of mutation would be increased if the effective energy of activation were lowered; this would be comparable to catalysis of a chemical reaction. As a matter of fact there is evidence that such catalysis is possible. Apparently there are genes that tend to speed up mutations in general, an effect similar to catalysis.[4] One might also expect that the rate of mutation could be increased by supplying the energy of activation, as in photochemical reactions. We have ample evidence today that radiation may bring about an increase in mutation rate. Ultraviolet radiation is an effective tool in this regard, and x-rays and other ionizing radiations also increase mutation rates. It has been shown that mutation rates may be increased by the application of chemical substances, notably compounds described as "nitrogen mustards." How these substances act we do not know, but the demonstration of chemical mutagens makes the treatment of mutation as a chemical reaction seem all the more reasonable, although it suggests that our scheme may be greatly oversimplified.

We might expect from our analogy that, while it should be possible to influence the rate of mutation, the direction of the mutations would as a rule be little affected. The proportion of forward and back mutations, which have been compared to the free energy change of a chemical reaction, should not be altered by speeding up the reaction either by lowering the energy of activation or by supplying the energy of activation. That is, we should not expect the kind of mutations to be affected by these factors, nor the proportion appearing in a population, although perhaps the analogy should not be carried too far in this respect. But if the gene is analogous to a polymer molecule, or a complex of such molecules, only certain variations in pattern should be possible and at the same time reproducible and stable—the conditions which were pointed out earlier in this chapter as being implicit in the concept of evolution by natural selection. These variations or mutations would not occur in a purely random fashion but would be restricted according to the configuration of the molecule or molecules

[4] Demerec, M., *Genetics* (1937) *22*, 469–478; Mampell, K., *Proc. Nat. Acad. Sci.* (U.S.) (1943) *29*, 137–144; *Genetics* (1945) *30*, 496–505; Ives, P. T., *Records Genetics Soc. Am.* (1949) No. 18, 96.

constituting the gene. This molecular configuration is dependent upon energy relationships and hence may be thought about in thermodynamic terms.

Whatever the nature of mutation, it will have to follow certain lines that are determined by molecular pattern and energetic relationships. Mutation, then, is not random, but may occur only within certain restricting limits and according to certain pathways determined by thermodynamic properties of the system. Thus, to state the case in a somewhat animistic fashion, the organism cannot fit itself to the environment by varying unrestrictedly in any direction. Certainly Darwin and many who have followed him have thought of variation as completely random, and hence providing all possible variety of handles for natural selection. In many ways this is still a good working point of view, but it may lead to difficulties if carried to extremes. The outcome of such ability to vary in all directions would be that, given time and opportunity, the organism might eventually vary in such a way as to fit itself to its environment with a very high, or indeed a perfect, degree of accuracy. Judgment of accuracy of fit is to a considerable extent subjective, and hence may be influenced by one's picture of the mechanism by which the fitting is accomplished. The tacit idea that mutation is completely random may thus lead to an overemphasis of the fit between organism and environment. Recognition that there are restrictions on mutation should introduce an appropriate skepticism into our thinking about the fitness of the organism to its environment.

No doubt some will find quite obvious the idea that there are restrictions on mutation, and see no need to labor this point. This would be true, perhaps, if the extent of the restrictions on the evolution of living systems which have been placed by various factors could be properly evaluated and incorporated into our thinking. But a complete evaluation of such restrictions is no doubt impossible to achieve in more than very general terms. Purely physical factors impose restrictions on the direction of organic evolution in a variety of ways, and must have imposed such restrictions from the very beginning, dating back at least as far as the appearance of life on the earth. To gain some comprehension of the nature and magnitude of these restrictions one would need to go back to the time of the origin of life, an area in which it might seem that unbridled speculation should be permitted. But even here speculation should be subject to restrictions. The next chapter is devoted to a consideration of this particular problem.

X · THE ORIGIN OF LIFE

"*When once you have taken the impossible into your calculations its possibilities become practically limitless.*"—SAKI

CERTAIN fundamental properties of modern living systems have been discussed in the last few chapters. How and when did such systems appear on the earth? Were the first of them endowed with all those properties that now seem essential to living organisms? Could these properties have appeared at different times, and what must have been their order of appearance? Did the manner of origin of the first living systems have great bearing on the later direction of evolution? These are questions which might be more easily answered by the congress of scientists in the time machine, pictured in an earlier chapter. Here one is limited to making what inferences he can, hoping that they may be reasonable and consistent. Perhaps it would be more profitable to study the questions themselves than to try to formulate answers.

The origin of life can be viewed properly only in the perspective of an almost inconceivable extent of time. A reasonably consistent fossil record extends back through the Cambrian or earliest of the Palaeozoic rocks to a time about half a billion years ago, according to accepted estimates. The Cambrian covers a period of the order of a hundred million years, at the end of which fossil representatives of most of the major groups of existing forms of life were present, although the chordates (the phylum including the vertebrates and man) and all higher plants were conspicuously absent. Even in the earliest of the Cambrian rocks a majority of the existing phyla are represented by forms which may be readily grouped alongside modern ones, and many of these were already quite advanced in their evolution. At least, they can hardly be described as simple, and one would guess—although there is no quantitative basis for doing so—that evolution from the simplest ancestors up to the complexity represented in the Cambrian rocks might have taken many millions of years. Going back beyond the Cambrian, the paleontological record is much less satisfactory. The pre-Cambrian extends back in time well over three billion years, and includes the oldest of the sedimentary rocks—the earliest in which fossils could be found. Fossils are rare in these rocks, which have undergone greater alteration subsequent to their formation than have younger ones. Probably many of the simpler organisms did not form hard parts, such as calcareous, siliceous, or chitinous skeletons, and hence have left no record. Objects from the pre-Cambrian once thought to be fossils have been shown to be inorganic in origin, and this led to considerable skepticism at one time; but the existence of fossil plants (algae) and some animal fossils has come to be generally conceded. The question is, how far back in the pre-Cambrian do living systems go. Improvements in radioactive dating and better identification of the fossil forms have now placed the earliest fossils much farther back in time than was thought possible a few years ago; recently fossil algae bearing rocks in South Africa have been assigned an age between 2.6 and 2.7 billion years,[1] about twice the greatest age suggested when this book was first written. The earliest sedimentary rocks have also been pushed back in time to about 3.8 billion years, increasing the pre-Cambrian by over a billion years. All such values are, of course, open to revision, but seem more likely to be increased than decreased. Accepting them for the present we may place the origin of life somewhere between 2.7 and 3.8 billion years ago, perhaps closer to the latter figure since the oldest identified fossils may represent relatively advanced forms.

In any case it would seem that there was lavish time for the initiation of the life process. The position taken here is that the transition from the non-living to the living took place in a series of steps spread over an unknown period of time, the length of which could, of course, depend on the definition of what constituted the earliest life; but so far as the time available is concerned, it would seem that the transition might have extended over many millions of years. A contrasting point of view, sometimes entertained, is that the vast sweep of time provided the opportunity for a single chance event which constituted the origin of life. These ideas will be considered more fully in Chapter XIII.

Discussions of the mode of origin of life always involve, sooner or later, questions of chance, and it may be well to orient ourselves with regard to such terms when referred to periods of time reckoned in millions or even billions of years. Let us return for this purpose to the model used in Chapter III, in which Mexican jumping beans were pictured as hopping about in two chambers of a dish separated by a

[1] Holmes, A., *Nature* (1954) *173*, 612–616.

barrier, as illustrated by the diagrams in Figure 2. According to the model, every so often a bean jumped over the barrier from one chamber to the other; this would be a chance event, the probability of which could be measured by its frequency of occurrence. In terms of our ordinary experience, if about every minute one of the beans jumped over the barrier we would no doubt think this a fairly frequent and hence "probable" event. If this happened only once in a year, on the other hand, we might not happen to be looking, and hence might say at the end of that time that such an occurrence was impossible—that it *never* occurred. But had we chanced on one occasion to see a bean jump over the barrier, we could no longer say that the event was impossible. We might still say that it was *improbable*, however. If once every hundred years a bean jumped clear out of the dish, we would have to consider this a still less probable event, or we might call it a more improbable one. The point to emphasize is that the longer the time period concerned, the greater chance there is for improbable things to happen.

Parallel to the vast amount of time for the occurrence of improbable events, there was a vast amount of material available for chemical reactions. In this mass relatively improbable local conditions may have come to exist from time to time, resulting in relatively improbable reactions. Referring to Table 2, one sees how small a portion of the nascent earth has contributed to the environment of living organisms. For example, the oceans constitute only about one-tenth of one per cent of all the atoms in the earth today. The amount of living material is more difficult to estimate, but the mass of the whole of existing organisms is probably about 10^{19} grams.[2] This is roughly a billionth part of the present mass of the earth, 5.974×10^{27} grams. Thus, regarded as a product of a great reaction system comprised by the nascent earth the original "yield" of living material would have been exceedingly small, even had it comprised the equivalent of that represented by all the living organisms on the earth today. We might think of living systems as a minute quantity of by-product from a complex chemical reaction carried out on a gigantic scale, a tiny mote in a great retort.

No matter how the problem of the origin of life is approached it seems necessary to admit that some events may have occurred which would appear highly improbable if viewed in our customary frame of experience. Because of unrecognized factors some of these happenings may have been less improbable, however, than they now appear. Certain events are always more probable than others. The combinations of atoms and molecules are always related to their thermodynamic

[2] Based on the estimate of V. I. Vernadsky, quoted by E. I. Rabinowitch, *Photosynthesis*, New York (1942) Interscience Publ.

properties, e.g. their free energies; for any set of conditions, the temperature, pressure, and concentrations of the various species of reactant molecules, etc., there will be a greater probability for some compounds to form than others. Thus, while referring to "chance" as having contributed to the origin of life—or having shaped subsequent evolutionary history—it should be recognized that the system under study is not a completely random one, but one in which some events are more probable than others. The cards are always dealt from a deck that is, to a greater or less extent, stacked. (See Chapter XIII.)

The Evolution of Chemical Complexity

Returning now to the evolution of the nonliving earth—already sketched in earlier chapters—emphasis may be placed upon certain factors leading to the chemical complexity that may be assumed to have existed on the earth's surface at the time of the appearance of life. Let us have recourse to a model rather than attempt to reconstruct exact happenings. Suppose, for this model, an isothermal system including a great variety of molecular species, among which there would be a large but limited number of possibilities of combination, determined by the free energy changes of the respective reactions. Now let us suppose something that could not possibly be true—that all the possible reactions proceeded at rates determined by their free energy changes, those reactions going most rapidly which entailed the greatest decrease in free energy. There would be a steady approach to the condition of least possible free energy of the system. This would represent a rigid following of time's arrow, with a rigidly determined outcome.

Rates of reactions are not, however, determined by their respective free energy changes, but, at ordinary temperatures, by their energies of activation, and there is no constant relationship between these quantities. If we are to consider our model system to be analogous to a real one, we must expect that some reactions will go much more rapidly than others. At one time certain reactions will be thermodynamically possible, but after an interval of time only certain ones of these will have taken place to an appreciable extent. At the end of the interval a different set of reactions will be possible than at the beginning, product molecules now being able to react with other product molecules or with some of those molecules present at the beginning which did not react in the interim. At the end of another interval the scene will again have changed, new types of molecule having appeared, and new possibilities of combination. Thus, in the course of movement in the direction of minimum free energy, great complexity of chemical compounds could appear, due largely to the existence of two independent factors: one, time's arrow, determining the general direction; and the other, rate of reaction, the detail of the pathways followed. In such a system each event would be a function of the events preceding; at no time should the process be thought of as completely random. There would be expected, too, a tendency for reactions once started in a given direction to continue in that direction, resulting in "channeling" along certain pathways to the neglect of others.

A further analogy may assist us here. If a large mound of earth were allowed to be eroded away by a constant flow of water over its surface, the water being applied just at the top of the mound, the force producing erosion would be proportional to the potential energy represented by the difference in level between the top of the mound and its base. Assuming purely mechanical factors, the rate of erosion would be proportional to this force. This potential may for our purposes be considered as analogous to chemical potential represented by free energy, although like most analogies it is not exact. As the mound decreased in height the potential energy would decrease in proportion and consequently the erosion would proceed more slowly. The geologist will recognize in this analogy the principle of peneplanation. The rate is not only determined by the potential energy represented by the height of the mound, however, but also by the refractoriness of the soil to erosion. The soil not being homogeneous, erosion might be expected to proceed more rapidly at some points of the mound than at others, resulting in the formation of channels in which the water would flow. Thus the harder parts would be relieved of the action of the eroding agent, and the tendency would be to increase the difference in rate of erosion between the hard and the soft parts. Thus, although we might be unable to predict where the channel would form, once formed it would tend to maintain itself rather than to seek new channels. Of course, if there were a stone visible in the mass of soil, we might predict that the soil would be eroded first rather than the stone, although we might not be able to predict exactly the course of the channel. [3]

In the light of this analogy, let us consider the directions of the various reactions that might occur in chemical evolution, and hence the kinds of molecules that might exist at any given time, in a reaction system such as we have pictured as a model. We could be sure that in the overall sense the reaction would go in the direction of minimum free energy—or if we dealt with an isolated system, maximum entropy —but we could only hope to make the most obvious guesses as to what particular combinations might occur. It seems inevitable that there would be some degree of channeling, leading to the predominance of certain types of compound over others. Yet it might be difficult to

[3] This analogy has been lifted almost verbatim from my first paper on this subject, *Am. Naturalist* (1935) *69*, 354–369

understand this predominance without knowing in detail the steps followed in the course of chemical evolution of the nonliving world. We could expect, as the result of this evolutionary process, a wide variety of chemical species, among which certain genera were more common than others.

The model reaction system used above is greatly oversimplified as compared to any that might have existed in the nascent earth. It was tacitly assumed to be a homogeneous system, the possibility of changes of state such as liquefaction from the vapor state and precipitation from solution being neglected. An isothermal system was specified, whereas at the time we envision the earth may have been cooling because it was losing heat by radiation to space faster than it was gaining energy from sunlight or other sources. The possibility of photochemical reactions brought about by sunlight was not mentioned—a very important factor after the advent of living systems, but one more difficult to evaluate in the evolution of a strictly nonliving world. Simple organic molecules may have been captured from space increasing the complexity, and there may have been other contributory factors;[4] but the interplay of all these factors would only add to the complexity and unpredictability of the evolutionary process, and they may be thought of as supplementary, or complementary, to those stressed in connection with the model.

There seems no question that a profusion of carbon compounds would have been formed in the great reaction system of the nascent earth. The types of these organic compounds would have been determined by thermodynamic and kinetic factors, and perhaps it would be best not to go very far in speculating as to their kinds, or the manner of their formation. The possibility may be mentioned, however, that some of the characteristic molecular patterns which occur repeatedly in living systems appeared in the course of this period of chemical evolution before the advent of living organisms,[5] and that perhaps chemical evolution was channeled along these lines in the way suggested above in connection with our model. We may ask ourselves whether in the great profusion of organic compounds we suppose to have been present, there might have been some as well fitted for biological processes as those that actually became incorporated into the patterns of living systems. No definite answer can be given because only the survivors, that is, those which have continued to be reproduced, are obvious to us. For instance, the question has been raised earlier whether the adenylic

[4] Kavenau, J. L., *Am. Naturalist* (1947) *81*, 161–184, has emphasized that the diurnal alterations of sunlight and temperature of the earth's surface may have been important factors in creating such special conditions.

[5] Such recurrent patterns will be discussed in Chapter XI.

acid system could not have been replaced by some other set of compounds not containing the element phosphorous. This system seems admirably fitted for its job, but the unique fitness of phosphorus for this role is certainly not as clearly established as that of hydrogen in the scheme of living systems as a whole. We only know that this special type of compound plays an apparently unique and essential role in living systems today.

At some time in this process of chemical evolution in a strictly non-living world, the first living systems appeared and from thence forward the course of that process was progressively changed, in certain aspects at least. This was over a billion years ago, perhaps very much earlier, and at that time there was no doubt already present a complex variety of carbon compounds. The difficulty of estimating the time of this event seems very great indeed, even if we could define satisfactorily what we mean by the "first" living organisms. The term natural selection is sometimes loosely applied to evolution in physical systems, and perhaps a basic difference between chemical evolution and evolution of living organisms should be pointed out. Let us think of the molecules of a given species, A, in the gaseous state. All the members of the species are not exactly alike since at any instant they have different amounts of energy. The energies of the individual molecules change continually although there is a mean energy characteristic of the conditions. Some of the molecules which achieve the necessary energy of activation may react with molecules of some other species, B, and so be "selected" out of the mass as a whole, a new species of molecule, AB, appearing. With changes in conditions (environment) other species may emerge. This kind of chemical evolution, rigidly governed by energetics and kinetics, we may think of as going on before the advent of life; but this is very different from mutation and natural selection in living systems. In the strictly chemical system molecules lack the property of self reproduction—the activated molecule does not perpetuate itself by reproducing its kind, but rapidly returns to a normal level if it does not undergo reaction. Reproduction of stable patterns and stable variants of these patterns is essential for evolution by natural selection. This was impossible before the advent of systems that could synthesize molecules having the stability and kind of variability associated with proteins and nucleic acids.

The Origin of Proteins and Nucleic Acids

Did the proteins and nucleic acids, the great importance of which was discussed in Chapter IX, come into being in a strictly non-living world? If so, there are reasons to think that the proteins, or at least long polypeptic chains, would have emerged first. Certainly it is likely that the monomers of which they are composed, the amino acids, appeared upon the earth before the nucleotides which are the monomers in nucleic acids. In 1953, two years after the first publication of this book, Stanley Miller carried out ingenious experiments which have had great influence upon subsequent thinking about the origin of life. Some years before, A. I. Oparin had presented evidence to indicate that the early atmosphere of the earth contained no oxygen, and about 1960 this idea was further supported by Harold Urey. Miller's experiment which was carried out in Urey's laboratory was designed to explore the possibility of chemical reactions taking place in the kind of atmosphere that might have obtained in the early years of the earth. A mixture of ammonia, hydrogen, methane and water vapor was exposed to an electrical spark with the result that organic compounds were formed; among these were the amino acids alanine and glycine. Whether the reactions were caused by the electrical discharge per se or the ultraviolet light emitted from it is not clear; if the latter, the wavelengths must be shorter than those reaching the earth with its present atmosphere (see Figure 4) since these are not absorbed by the substances concerned. The experiment has since been repeated under somewhat different conditions, and the possibility that the simpler amino acids could have formed in the early atmosphere of the earth seems clearly established.

But the question arises as to how these amino acids could have become joined together into polypeptide chains. It is commonly assumed today that life arose in the oceans, J. B. S. Haldane's "dilute hot soup" providing a supposedly appropriate medium. But even if this soup contained a goodly concentration of amino acids, the chances of their forming spontaneously into long chains would seem remote. Other things being equal, a dilute hot soup would seem a most unlikely place for the first polypeptides to appear. As we have seen, the free energy change for formation of the peptide bond is such that at equilibrium about one per cent of the amino acids would be joined together as dipeptides, granting the presence of appropriate catalysts. The chances of forming tripeptides would be about one-hundredth that of forming dipeptides, and the probability of forming a polypeptide of only ten amino acid units would be something like 10^{-20}. The spontaneous formation of a polypeptide of the size of the smallest known proteins seems beyond all probability. This calculation alone presents serious objection to the idea that all living systems are descended from a single protein molecule, which was formed as a "chance" act—a view that has been frequently entertained.

Of course, one may imagine that the first polypeptides were formed in relatively non-aqueous media. It has long been laboratory practice to carry out such reactions in solvents other than water. Oparin suggests that life originated in coacervates, aggregates of molecules forming a phase discontinuous with a surrounding aqueous medium; but we have to ask the nature of the coacervates and how they became formed. The experiments of Sidney Fox, in which he has produced polypeptides under a variety of conditions, involving heat and relative dryness, are particularly intriguing. One might picture polypeptides forming early in earth's history in dried up puddles. I have even suggested (1961), perhaps somewhat facetiously, that such a puddle, heated by sun and undergoing intermittent heating and cooling as the earth rotated, might, by providing for successive formation and dissolving of polypeptides, have been the basis for the first replicating systems. In any case, a medium relatively low in water, rather than a dilute solution, would be a more probable place for polypeptides to form; but such conditions assume the existence of dry land, so such a process could only have occurred after at least limited land masses had appeared on the earth.

The joining together of nucleotides to form polynucleotides involves the removal of water, in a manner similar to that in the formation of polypeptides. The forming of the nucleotides themselves presents other problems. For example, the element phosphorus is a necessary component and this element would not have been available to any extent until after land masses had formed and been eroded for a long time. This may be a basic reason for assuming that polypeptides to be older on the earth than polynucleotides.

Let us consider briefly another intriguing problem with regard to the origin of the proteins, or rather the origin of their building blocks the amino acids. All the amino acids derivable from proteins are of the "left-handed" variety. Amino acids synthesized in the laboratory are a mixture of the right- and left-handed forms, and thermodynamically the two are indistinguishable. Yet the living organism uses only one kind in building proteins. Why should only left-handed amino acids be utilizable? We have no very good answer to this question unless it is that the patterns or templates upon which proteins are reproduced are in some way essentially left-handed. Possibly the stability of protein structure, or alternatively its reproducibility in the living organism, is linked with the existence of only one kind of optical isomerism in the units from which it is built. If so, the existence of only one kind of isomer in the amino acid units of the proteins would have the most fundamental importance. But this does not help to answer why only the left-handed isomer is formed. It might have to be either the left-handed one only or the right-handed one only, but why the one rather than the other? Whether the existence of only one kind of isomer is fundamental to protein structure or not, it seems necessary to associate this limitation of pattern with the mode of origin of protein molecules. Was it a chance event related to the formation of the first protein that dictated the left-handed pattern? Or was there some factor which caused left-handed amino acids to predominate during the time the first protein molecules were being formed? Both solutions to the problem have been proposed[6] and different mechanisms suggested, but we have no definite answer.

However one tries to picture the emergence of these all-important polymer constituents of living systems, one has to assume various special conditions, and if he should try to specify all these conditions in order to set up experiments to simulate them he would be faced with a corresponding number of choices. And each choice made by the experimenter might be construed as contributing to the improbability of the origin of life on this earth. But this is something it will be better to take up again in Chapter XIII. Here it may only be pointed out that the simulation of the conditions in each choice would involve a certain amount of work on the part of the experimenter, and this brings us to the problem of the mobilization of the free energy that must have been required to shift equilibria toward the formation of these polymers, and for the many other processes involved in the earliest replicating systems.

The Mobilization of Free Energy

Few of those concerned with the problem of the origin of life seem to have given more than passing attention to the question of mobilization of free energy for the reproduction of the original living systems.[7] Since the reproduction of proteins could not have gone on without a means of energy mobilization, it might almost be necessary to assume that these two processes had their origin at the same time, unless indeed the latter actually antedated the former. In all modern organisms energy metabolism is so closely dependent upon the existence of proteins, catalysis by enzymes being an intimate part, that it is difficult to see how they could have evolved separately.

At any rate, the problem of energy supply for the first living organisms seems fundamental, and we must make some shift to attack it.

[6] For discussions see: Eyring, H., Johnson, F. H., and Gensler, R. L., *J. Physical Chem.* (1946) *50*, 453–464; Gause, G. F., *Optical activity and Living Matter*, Normandy, Missouri (1941) Biodynamica; Wherry, E. T., *Proc. Pennsylvania Academy of Science* (1936) *10*, 12–15; Dauvillier, A., and Desguin, E., *La genèse de la vie*, Paris (1942) Hermann.

[7] The discussion by A. Dauvillier and E. Desguin, *loc. cit.*, is an exception.

Since there is no direct correlation between free energy and rate of reaction, many energy-rich compounds could have been left behind in the general movement toward greater entropy. Presumably some of these could have been utilized as sources of free energy for the synthesis of the first proteins, whether the latter were formed within or outside living systems. The compound glucose may be chosen by way of illustration. Although readily oxidizable by living organisms with the release of large amounts of free energy, at room temperature this compound may in their absence remain unoxidized almost indefinitely. Living organisms are able to accelerate this reaction by means of enzymes, but, in the absence of these catalyzing systems, chemical potential might be retained in such a compound for a long time. While interest tends to focus on the carbon compounds, particularly carbohydrates and other substances concerned in the energy metabolism of modern organisms, it should not be forgotten that there could also have been considerable sources of chemical potential in the inorganic world, and that there are living organisms that can use inorganic reactions as the sole source of their energy supply.

Another possible source of energy-rich compounds lies in photochemical reactions which may have brought about the storage of the energy of sunlight during the period of chemical evolution. The idea that processes similar to green plant photosynthesis went on in the nonliving world dates back to the nineteenth century, and is still current. About 1870 Baeyer proposed that photosynthesis involves the formation of formaldehyde, CH_2O, from carbon-dioxide and water. The reaction may be written

$$CO_2 + H_2O = CH_2O + O_2; \qquad (X\text{-}1)$$
$$\Delta F = +\ 125 \text{ kg-cal/mol;}$$
$$\Delta H = +\ 134 \text{ kg-cal/mol.}$$

Formaldehyde was then supposed to undergo polymerization to form carbohydrates. There is now conclusive evidence that the above reaction is not concerned in green plant photosynthesis in spite of its general resemblance to reaction VII-20 which characterizes that process in an overall sense. But the idea persists that during the early history of the earth, reaction X-1 might have been brought about by the action of ultraviolet radiation without the intermediacy of living organisms. The quanta in sunlight are inadequate to supply the energy necessary to forward this endergonic reaction, however, and the difficulty of summing quanta in simple photochemical reactions has already been discussed. The smallest quanta that could supply the energy for this reaction correspond to about wavelength 0.18 μ, which is much shorter than any found in sunlight today.[8] Whether radiation of such short wavelengths ever reached the surface of the earth in sufficient quantity to make this reaction an important source of chemical potential is open to serious question. A. Dauvillier and E. Desguin suggest that short wavelength radiation from the sun could have supplied the requisite quanta in the absence of an oxygen atmosphere. It seems doubtful, however, that this source could have contributed importantly to the storage of energy rich compounds. No appreciable amount of such radiation penetrates to the surface of the earth today; whatever small amount reaches the outer atmosphere is absorbed by the oxygen, and the thin layer of ozone which effectively cuts off radiation of wavelengths shorter than about 0.29 μ. For effective penetration of shorter wavelengths it would have to be supposed that the atmosphere was virtually devoid of oxygen, and, as a result, free of ozone, which is formed from oxygen by the shorter wavelengths of sunlight. The idea that there was an anaerobic period in the evolution of the earth is one of long standing which has now received strong support; in this case shorter wavelengths could have reached the earth. But even a small quantity of oxygen—and reaction X-1 would produce this substance—would eliminate the shorter wavelengths of sunlight. Even granted the penetration of these short wavelengths to the earth's surface, there remains the fact that they would not be strongly absorbed by water or CO_2.

But perhaps we should not be too dismayed by the inadequacy of reaction X-1 as a source of the energy-rich compounds to be used by the first living organisms. The reaction

$$CO_2 + 2H_2S = CH_2O + 2S + H_2O; \quad \Delta F = +19 \text{ kg-cal/mol}; \quad \Delta H = +22 \text{ kg-cal/mol}. \qquad \text{(X-2)}$$

which is similar to the overall reaction of photosynthesis by the green sulfur bacteria (VII-22), is somewhat more plausible, since it would require much smaller quanta, and these could be supplied in great quantity by sunlight that comes through our modern atmosphere. This reaction is not known to be brought about by the action of sunlight, however, and it is only offered by way of illustration of the point that, considered from a thermodynamic standpoint only, a large number of photochemical reactions might be imagined that could have served to store the energy of sunlight in various organic or inorganic compounds during the period of chemical evolution. On the other hand, the fact that a photochemical reaction may be thermody-

[8] I discussed this difficulty in more detail some years ago, *Am. Naturalist* (1937) *71*, 350–362.

namically possible does not necessarily mean that it will occur, for some very special mechanism may be needed, such as exists, for example, in modern photosynthesis. Reactions such as X-2 suggest, however, that we need not restrict our thinking about photochemical processes for the storing of the energy of sunlight to those taking place only in the short wavelength region. In general, organic compounds are likely to absorb in this region, and of course the shorter the wavelength the larger the amount of energy that can be introduced into the absorbing molecule by the corresponding quantum. To be sure, absorption of wavelengths shorter than about 0.32 μ by substances in the living cell results in destructive action which is apt to be thought of as a special kind of action of ultraviolet radiation. But the idea that ultraviolet radiation, as such, is endowed with some special unique power to forward photochemical reactions is a residue from a time before the introduction of quantum theory changed the whole aspect of photochemistry.

But is it necessary, after all, that the exergonic reactions supplying energy for the synthesis of the first proteins should have involved such high free energy changes as does, for example, the oxidation of carbohydrate? The minimum energy requirement for the formation of a peptide linkage is not high, and could be supplied by the transfer of a single energy-rich phosphate bond, if there were any around.

The Metabolism of the "First" Living Organisms

In a recent article Garrett Hardin[9] says, "Regarding the origin of life there are at the present time only two scientific hypotheses that are given serious consideration. . . . we may call these the 'Autotroph Hypothesis' and the 'Heterotroph Hypothesis.' According to the former, the first form of terrestrial life must have been some organism that could manufacture its own organic substance out of inorganic substrates, as can contemporary green plants. The Heterotroph Hypothesis, on the other hand, states that the first 'organism' was one of severely limited synthetic abilities, subsisting on a readily available menu of organic materials formed by nonvital processes." He goes on to say that only the heterotroph hypothesis, which has been developed of recent years by A. I. Oparin, J. B. S. Haldane, A. Dauvillier and E. Desguin, N. H. Horowitz and C. B. van Niel merits serious consideration. I must admit to a certain degree of uncertainty as to just how these two hypotheses are to be distinguished. If the autotroph hypothesis assumes that the first living organisms sprang directly from inorganic matter, whereas the heterotroph hypothesis assumes that it was necessary that there first be present a complex system of organic compounds, then I must unhesitatingly subscribe to the latter. If the

[9] *Scientific Monthly* (1950) *70*, 178–179.

two hypotheses are to be separated on the basis of whether the first living systems enjoyed on the one hand autotrophic or on the other hand heterotrophic metabolism, as they exist in modern organisms, then I find greater difficulty in choosing between the two.

To me, the greatest problem regarding the origin of life lies at another level. In the first place, it seems necessary to face the difficulty of deciding what was the first organism. The origin of life represents a transition from the nonliving to the living, which I have great difficulty in imagining as a sharp one. I do not see, for example, how proteins could have leapt suddenly into being. Yet both heterotrophic and autotrophic metabolism are, in modern organisms, strictly dependent upon the existence of proteins in the form of catalysts. The riddle seems to be: *How, when no life existed, did substances come into being which today are absolutely essential to living systems yet which can only be formed by those systems?* It seems begging the question to suggest that the first protein molecules were formed by some more primitive "nonprotein living system," for it still remains to define and account for the origin of that system.

Although they do not help us to solve the dilemma just posed, certain ideas associated with the heterotroph hypothesis help greatly in thinking about the early evolution of living systems. Oparin suggested some years ago that the first living organisms on the earth exhausted the supply of organic materials which were essential for the origin of the first life. Thus spontaneous generation of life could not again occur because the requisite materials were no longer available. Horowitz,[10] extends this idea, assuming that there were present to being with complex organic compounds—presumably formed in the process of chemical evolution—and that the first organisms utilized some of these compounds which they were unable to synthesize. When the initial supply of these essential substances was exhausted, the original organisms perished. But in the meantime there had occurred mutations resulting in living forms that could carry out the synthesis of the essential complex compounds from simpler ones remaining in the environment. Further mutations provided additional steps in the synthesis, until finally there were organisms which could synthesize all of the essential complex compounds. The intermediate mutant forms, unable to perform the complete synthesis, tended to be eliminated by natural selection as intermediate substances were exhausted from the environment. Thus the evolution of the synthesis of complex compounds which were initially provided by the environment is nicely accounted for on the basis of mutation and natural selection. I must point out, however, that Horowitz's hypothesis still leaves a seemingly unbridged

[10] *Proc. Nat. Acad. Sci.* (U.S.) (1945) *31*, 153–157.

gap in the story of the origin of life. For does not the invoking of natural selection postulate the prior existence of that for which the origin is sought? Natural selection itself seems only possible in systems having a complexity corresponding to at least that of the proteins. What were the evolutionary steps which antedated the origin of such systems? Who would venture much more than to suggest that time's arrow played an important role?

But let us return to the question of the earliest type of metabolism. Assuming that the original living organisms were heterotrophic, they had, sooner or later, to evolve some form of photosynthesis if their energy requirements were to be satisfied. It seems probable that they would have exhausted their supply of energy-rich compounds within a relatively short time. We may make some calculations regarding this point, although these cannot be expected to give an accurate idea of the situation at the time of the origin of life. At present, photosynthetic plants fix each year about 1.5×10^{16} moles of CO_2 as energy-rich compounds of the carbohydrate type. Assuming a balance between energy gathering and energy spending, about the same amount of energy-rich carbon compounds is returned to CO_2 in the same length of time. Hence it may be assumed that living organisms use collectively about 1.5×10^{16} moles of energy-rich carbon compounds of the carbohydrate type per year. There are about 5×10^{16} moles of carbon in the atmosphere as CO_2. If this is taken to represent the initial amount of carbon available, we can see that the supply would be exhausted in a few years at the present rate of metabolism by living organisms, even if it had all been present as carbohydrate to begin with. Of course there may have been a much greater supply of carbon available, and we may try some further estimates. Assuming that all the O_2 of the atmosphere has originated from the reduction of CO_2, a corresponding amount of carbon might once have been available. This would amount to about 4×10^{19} moles, which would last at our present rate of metabolism only about three thousand years. If all the carbon in the earth—estimated to be about 2.4×10^{23} moles—had once been available as carbohydrate it would have lasted about sixteen million years, but we hardly need to entertain this as a possibility. Of course it is ridiculous to suppose that the first living organisms required more than a tiny fraction of what is used at present, but on the other hand it is ridiculous to assume that more than a small fraction of the available carbon was present as energy-rich compounds useable by living systems at the time of their origin. There would seem to be no way of replenishing the supply of such compounds except by capturing energy of sunlight by means of some photosynthetic process.

If the first organisms could have gained their energy from inorganic sources, the supply might have been greater, but we have even less basis for such estimates. Certainly[11] the supply was not inexhaustible, and sooner or later some mechanism for capturing the energy of sunlight had to be evolved, if life were to continue. However we regard the problem, we must admit that photosynthesis of some kind, perhaps very different from any we know today, arose very early in the course of organic evolution, if indeed it was not involved from the beginning.

Spontaneous Generation

That life was "spontaneously" generated from nonliving matter at some time in the very remote past, and that this process has not been repeated for a very long time, are two basic tenets accepted by the great majority of biologists. The first idea has been followed in all the foregoing discussion in this chapter, and seems a necessary part of the evolutionary concept. The second idea is supported by a multitude of experiments, and has not been seriously questioned since the time of Pasteur. But the experimental proof may not be as convincing to others as it is to biologists. I was not so long ago rather startled by the assurance with which a very competent physical chemist told me that spontaneous generation is no doubt going on all the time, but that we fail to observe it. This, I suppose, must be construed to mean that the experiments that have been done do not demonstrate conclusively that spontaneous generation does not occur. And as a matter of fact, this might be hard to disprove. But when one thinks of it, the best proof of the nonoccurrence of spontaneous generation within the period of the paleontological record comes from another source. It is probably the intuitive recognition of the incompatibility of repeatedly continued generation of primitive life with evolutionary history that satisfies the modern biologist in this regard, as much as the innumerable failures to obtain cultures of living organisms or of viruses from sterile media. For if spontaneous generation occurred with any degree of frequency, the whole evolutionary picture could be continually repeating itself before our eyes. In fact, it might be difficult to find evidence of evolution, for we would witness all its stages, rather than see an existing set of diverse forms tracing back toward a common ancestry through a series of extinct relations. I think the biologist is going to cling to the idea that spontaneous generation has not occurred during at least the last billion years; for he must demand, before he abandons

[11] It is suggested from time to time that the original living systems were autotrophic iron bacteria, which effected the deposition of some of the great iron deposits of the world. But these deposits could have been formed by purely chemical means (Harder, E. C., *Iron depositing bacteria and their geologic relations*, U.S. Geol. Survey, Professional Paper No. 113 (1919)). It is doubtful, too, that the iron bacteria are truly autotrophic.

this position, not only a demonstration of spontaneous generation under existing conditions, but an explanation of how the observed evolutionary relationships could have come about if there had been a continuous rebirth and at the same time evolution from primitive forms.

But in adopting this point of view the biologist faces another problem. How was it possible for life to have been generated at one time on the earth, yet this event not be again repeated during at least the past billion years of the earth's lifetime? A general answer is that the conditions no longer exist which once made the spontaneous generation of life possible. But what was the nature of the critical change or changes in these conditions? Oparin suggested that under existing conditions incipient new life or its building materials—say the necessary quantities of amino acids—are destroyed before new life has an opportunity to develop, because of the "predatory" activity of already existing life.[12] With the profuse development and distribution of living systems over the surface of the earth, there may exist no longer a nook or cranny where such incipient life could gain a foothold. Perhaps this is as good an answer as is to be hoped for, and it would be difficult to deny that this factor might be very important in preventing the continuance of spontaneous generation. One may suspect, however, that there were other factors. Perhaps there were chance happenings that occurred only once during the period when life was arising. The nonliving world has also evolved, and critical conditions that were essential to the origin of life may no longer exist on our planet.

An idea that has intrigued many, but which at best only pushes the problem back one step farther, is that minute spores were carried to the earth, "seeding" it with life. This hypothesis was given serious consideration by Svante Arrhenius,[13] who calculated that such spores could have been carried through space as a result of radiation pressure. Arrhenius recognized that the spores would have been subjected to a long intense bath of ultraviolet radiation within the reaches of our planetary system, but he disposed of this objection by saying that such radiation would not be destructive in the absence of atmospheric oxygen. It was thought in his day that O_2 was involved in the destructive effects of ultraviolet radiation on living systems, but it is now known that these are not diminished by reduction of oxygen pressure to low levels. The heating of the spores by absorbed radiation offers another objection. Since we now know that at least simple carbon compounds exist in outer space, it is conceivable that some key compounds may

[12] G. Hardin in a paper entitled "Darwin and the heterotroph hypothesis," (*Scientific Monthly* (1950) *70*, 178–179) quotes a passage from a letter of Charles Darwin which contains essentially the same idea.

[13] Arrhenius, S., *Worlds in the Making*, New York (1908) Harper and Bros.

have been "seeded" on the earth from that source, but this is a much different thing from the transplantation of organized life from other worlds. The general idea that life was transplanted to the earth from some other region of the universe can probably be safely disregarded. And this brings us again to the position that life was generated spontaneously on the surface of the earth—whether repeatedly or as a single event, or continuously over a considerable period—at some time prior to a billion years ago, but that life has not been spontaneously generated for at least that long a time, unless it was almost immediately wiped out by the activity of existing living organisms.

Some Relationships

All speculation regarding the origin of life is, of course, essentially an extrapolation. Starting with what we know about living organisms as they exist today, we proceed backward in time through the fossil record, into a region about which there is no information that can be regarded as in any way exact. The extrapolation depends largely upon apparent relationships, which permit a latitude of interpretation. In order to summarize some of these, Figure 20 may be useful, but the reader should be forewarned that it is not intended to represent an "evolutionary tree" in any detailed sense. He is at liberty to rearrange it as he pleases, but it is hoped that he will first follow it as a guide in the summary of the present chapter. No quantitative interpretation is to be placed upon the vertical distances, which are supposed to indicate the order of certain events in time, but not the actual times involved. The only intimation of a definite dating is the rough placing of the beginning of the Cambrian, which has been spread vertically to indicate the lack of exactness of timing of the events referred to it.

At the top of the diagram the "origin of life" is represented as spread over a considerable length of time, a period during which certain essential things happened, rather than as a single event. During this period some means of energy mobilization appeared, and the reproduction of complex molecules became possible. Which of these came first the diagram is not supposed to indicate, the arrangement there being one of convenience.

Neither heterotrophic or chemosynthetic organisms could have lasted alone on the earth, and photosynthesis must have appeared very early. In the diagram "primitive photosynthetic organisms" are shown emerging from the period of the "origin of life," with the blue-green algae appearing somewhat later. I do not mean to imply that the blue-greens are in direct line of descent to the higher plants, but a common ancestry is suggested. Paleontological evidence indicates that the blue-green algae are very ancient, it being generally agreed that calcareous deposits found in pre-Cambrian rocks are of a type made by representatives of this group living today. The modern blue-greens carry on the same type of photosynthesis as do the higher plants, and it seems reasonable to consider them as very early members

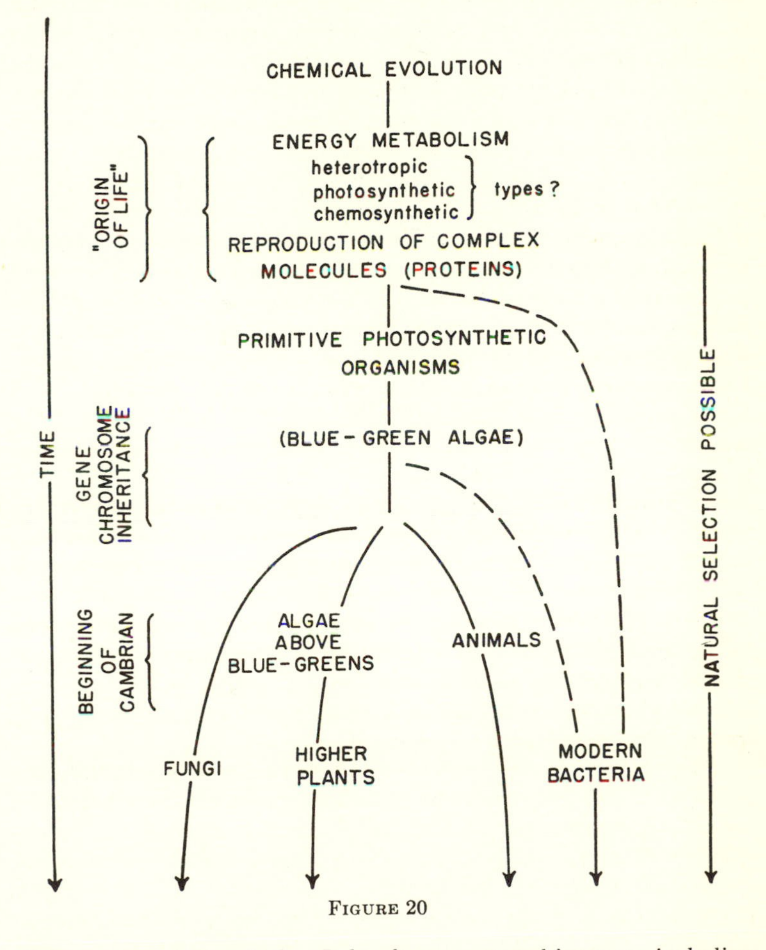

FIGURE 20

of the same common stock. It has been suggested by some, including at one time myself,[14] that the bacterial photosyntheses represent primitive types from which modern green plant photosynthesis is descended;

[14] *Am. Naturalist* (1937) *71*, 350–362. The scheme for the evolution of photosynthesis elaborated therein was based on comparative energetics and postulated a serial development from one to four quantum photosynthesis. It should now be relegated to limbo, having served its purpose, if any.

but it can be as easily contended that the bacterial forms are later developments. Both possibilities are indicated by dotted lines in the diagram.

A common ancestry for the green plants, fungi, and animals is suggested by the similarity of their nuclei and hereditary mechanisms. Definite nuclei and chromosome systems are demonstrable in the fungi and in the algae other than the blue-greens, as well as in the animals and higher plants. The fungi and animals are heterotrophic, while the algae and higher plants are photosynthetic. It is simplest to think of their common ancestor as photosynthetic, and their descent from primitive photosynthetic forms is indicated in the diagram. The gene-chromosome mechanism of inheritance common to the above forms seems to be absent in the blue-green algae, although structures resembling chromosomes have been described.[15] They have been placed in the diagram before the development of such a system. The bacteria lack definite chromosomes, and have been indicated in the diagram as arising either before or after the blue-green algae. Perhaps the bacteria, which contain forms having a wide variety of metabolisms including heterotrophic, chemosynthetic, and photosynthetic, may have a complex origin and should be indicated as branching off at more than one place in the diagram. Because of their apparent simplicity, they are generally regarded as primitive, but they have left no definite record in the earlier rocks. By the beginning of the Cambrian, algae above the blue-greens and animals of considerable complexity are present, but the higher plants do not appear until later. The position of the supposed common ancestor of these forms has been only vaguely indicated in the figure. It is intriguing to think that it was a one-celled flagellate organism such as we know today in both photosynthetic and heterotrophic forms, making them subject to claim by both the botanist and the zoologist.

There can be no doubt that the production of new species is and has been closely linked with the hereditary machinery represented by the gene-chromosome system of modern forms. Most of the modern concepts of evolution involve a genetic approach, the mutation and recombination of genes providing the handles for natural selection. But examination of the intricacies of the process of mitosis alone, as illustrated in Plates III and IV, gives the impression that this is not a simple or "primitive" machinery. In forms that are regarded as primitive, such as blue-green algae and bacteria, nuclear structure, if it exists, is certainly less complex. It seems reasonable to assume that the mechanism of inheritance in the earliest living organisms was considerably simpler, and that the complex chromosome and gene mechanism

[15] Spearing, J. K., Arch. f. Protistenkunde (1937), *89*, 209–278.

found in the modern higher plants and animals is a later development. The presence of the same kind of complex hereditary mechanism in both kingdoms of the living world is perhaps one of the best evidences of their common ancestry, and there seems every reason to think that in their beginnings, and perhaps for a long time after, living systems possessed a much simpler kind of hereditary machinery. The basis for evolution by natural selection should have been laid as soon as there was the possibility of persistent reproducible variations in pattern, and this could have been possible only after the advent of large complex polymer molecules. But this may have been a long time before there arose anything localized enough to call a gene in the classic sense. Natural selection is indicated in the diagram as being possible as soon as systems arose in which complex molecules could be reproduced. It is reasonable to think that these were nucleoproteins. Such systems are indicated as occurring before the rise of the gene-chromosome mechanism of heredity; perhaps they should be placed a very long time before.

In the introductory chapter I suggested that different scientists might disagree radically as to the moment of origin of life on the earth—even if they could, by some magic, witness the succession of events that actually did occur. This suggestion was based upon the difficulty of deciding what constitutes life, and the likelihood that the criterion for its origin might be influenced by the background of the individual scientist. A number of major properties are essential to living systems as we see them today, the origin of any one of which from a "random" system is difficult enough to conceive, let alone the simultaneous origin of all. As examples I would suggest: the origin of complex polymer molecules; the origin of mobilization and transport of free energy; the origin of photosynthesis; the origin of the gene; the origin of the cell. The list might be extended, and it is obvious that among those named some might stimulate the interest of one scientist more than another. But suppose we abandon the idea of a definite moment of origin and assume that a series of events represents the beginning of life rather than one definite point in this series. How can we judge as to the sequence of these events? For example, although it may be obvious that the gene could not have arisen before there existed complex molecules, how can it be decided whether or not such molecules arose before the initiation of life itself? We may have to remain without an answer to such questions, lacking as we do a time machine.

XI · IRREVERSIBILITY AND DIRECTION IN EVOLUTION

"*. . . our sense of the forward movement of time and the law of increase of entropy are based upon or grow out of the same fundamental conditions in nature.*"—W. S. FRANKLIN

ONE thinks of evolution as a series of events in time; of steps that have not been retraced, unless perhaps during short intervals and in a restricted sense only. The idea of the irreversibility of evolution is often spoken of as Dollo's law, after the Belgian paleontologist who seems to have been the first to point out the evidence of this in the fossil record.[1] But actually, the nonrecurrence of experienced events may be one of the oldest notions of the human mind, for in any real experience our sensation of time is unidirectional, and the irreversibility of history and of evolution seem to be corollaries of this. Why, then, is the question still debated? Probably because there is sought an intimate mechanism to account for the irreversibility of evolution as applied to its individual "steps." We may possibly encounter some small difficulty in accounting for irreversibility at this level, but in an overall sense evolution can hardly be thought of as other than irreversible. Being a one-way process in time, evolution appears to have a direction as we view it from the present, looking back over the course of millions of years. Its flow may be pictured as that in a river delta, where the main stream breaks up into numerous channels and these divide again into smaller ones. So we think of the primitive stream of living things having divided into a variety of evolutionary courses increasing in number as time went on. The existing species may be compared to the mouths of the various channels of the delta where they approach sea level; their evolutionary relationships trace back to the

[1] Dollo, L., *Soc. Belge. Geol. Bull.* (1893) *7*, 164–166; and see account by Petronievics, B., *Science Progress* (1919) *13*, 406–419, English translation in *Smithsonian Report for 1918*, pp. 429–440.

main stream.

The idea of direction in evolution is sometimes associated with the term *orthogenesis*, a word whose usage seems to be somewhat uncertain. If employed to describe the persistence of evolution in certain pathways, the term may be useful.[2] On the other hand, if an extraphysical directing factor is implied—as seems often to be the case when mention is made of a "theory of orthogenesis"—only confusion results.[3] Sometimes in connection with its use in the latter connotation, evidence of supposedly nonadaptive "orthogenetic" series which are thought not to be explainable in terms of natural selection is cited. Increase in size of the horns of successive species of titanotheres, or the progressive elongation of the canine teeth of the saber tooth tiger are examples frequently used. But as G. G. Simpson[4] so clearly points out, if properly interpreted there is nothing in such evidence which cannot be as readily explained in terms of modern ideas of evolution by natural selection. Under the circumstances it might seem best to avoid the word orthogenesis altogether, yet other terms that may be used to describe the same aspects of evolution—direction, pathway, trend—also lack rigid physical definition,[5] although the things to which they are applied seem real enough to have attracted the attention of all evolutionists.

Homology and Analogy: Recurring Patterns

Perhaps the most convincing evidence that evolution has been channeled along certain paths, giving the impression of direction, is derived from homology and analogy. The first term refers to the developmental derivation of a given part, and the latter to its function. The wings of a bird and the arms of a man are *homologous*, since they derive from the same basic structures in the appendage pattern of the vertebrates. The wings of a bird and those of an insect are *analogous* in terms of the function they perform. The ability to fly has been devel-

[2] E.g., Lwoff, A., in *L'evolution physiologique*, and Florkin, M., in *L'evolution biochimique*, introduce respectively the terms *orthogenèse physiologique* and *orthogenèse biochimique*, which both authors seem to employ in the descriptive sense.

[3] Jepsen, G. L., *Proc. Am. Philosophical Soc.* (1949) *93*, 479–500, in reviewing a large number of papers in which the term appears, finds orthogenesis used in both the descriptive and causational sense; paleontologists as a rule employing it descriptively, botanists and zoologists more often with causational connotation.

[4] *The Meaning of Evolution*, New Haven (1949) Yale University Press.

[5] There seems danger, here, of being caught in a tangle of vague terminology from which escape may be difficult. The phrase, "evolution in a *straight line*," is a case in point; this straight line has no apparent physical dimension, and to me the term lacks meaning and seems unfortunately chosen.

oped in these two cases along very different lines, the appendage pattern in the insects being radically different from that of the vertebrates. Presumably, natural selection has directed along very different pathways the formation of appendages that carry out the same advantageous function. Evidence from comparative morphology (in both past and existing forms) and from embryology shows a very clear evolutionary direction in these two cases. The whole picture of evolution is one in which derivation of new patterns comes from modification of those existing—never by the return to a former starting place for a new "try," unless in some very minor instances which we may neglect in our overall view. Whereas the mediaeval and renaissance artists found no restriction to hinder their creation of hexapod mammals, Nature has been rigidly limited in this respect, since she has been able to follow only certain established channels.

Homology and analogy are most obvious in morphological structures, but perhaps they may be more clearly viewed at the chemical level. The field of comparative biochemistry, thanks to the work of Ernest Baldwin, Marcel Florkin, A. J. Kluyver, André Lwoff, C. B. van Niel and a host of others, has now reached a point where there is a wealth of material from which to choose examples. The total number of organic compounds known is tremendous, and the number that might be synthesized in the laboratory, given time, materials, and desire, stretches toward infinity. Yet among all these possibilities only certain ones are reproduced by living organisms, and these appear to be built up according to a strictly limited number of structural types. If we examine these limitations and search for the origin of chemical types, we may find, as Florkin has pointed out, that biochemical homology is more fundamental than morphological homology, and may require a more basic approach. The fact that biochemical compounds can be arranged in relatively few series of homologous structural catagories, just as living organisms can be so classified, seems evidence enough that there is a general limitation to certain basic patterns. But as in the case of morphological structures, patterns that are not at all homologous may play analogous roles in the performance of specific physiologic functions.

It would be beyond the scope of this book to attempt a complete classification of the chemical structures that are found in living systems, and the details of their recurrence in association with widely different biological processes having no seeming relationship. The tetrapyrrole ring, illustrated by examples in Table 9, seems a particularly suitable choice for discussion. This structure is present in chlorophyll, the principal light absorbing pigment in green plant photosynthesis, in hemoglobin and some other respiratory pigments of animals, and again in enzymes concerned in biological oxidations in both plants and animals. Then, after these important roles, it turns up sporadically in lesser ones—as a coloring pigment in a molluscan shell, and in the feathers of a few isolated species of birds. Sometimes compounds of this type exert injurious effects by rendering organisms susceptible to injury by light.[6] Here is a type of chemical structure which Nature seems unable to escape. It appears, if an animistic metaphor may be permitted, as though having become intrigued with this pattern, she returns to it over and over again for lack of imagination.

In seeking a more mechanistic explanation, let us examine the apparent uniqueness of these substances for the positions they occupy. Can we believe that the tetrapyrrole structure is so exactly fitted for all the jobs it does, that no other could possibly replace it in these specific roles? We may find argument against this point of view in the case of the respiratory pigments. These substances, all of which are conjugated proteins, are involved in the transport of oxygen to the tissues of animals having more or less well-developed circulatory systems, the ability to hold oxygen in loose combination being associated with the prosthetic groups. In the *hemoglobins* of the vertebrates, the tetrapyrrole structure *protoheme* is combined with globin, the protein part of the molecule, in the ratio four to one. In the "invertebrate" hemoglobins, or *erythrocruorins*, the same heme is present but is combined with globin in other proportions. The *chlorocruorins*, green respiratory pigments found also among invertebrates, contain a slightly different tetrapyrrole structure, *chlorocruoroheme* (Table 9). All these compounds have an atom of iron in the tetrapyrrole ring, but there is yet another type of invertebrate respiratory pigment, *hemerythrin*, which contains iron but no heme. A further departure is *hemocyanin*, a green copper-containing protein which has neither iron nor tetrapyrrole ring. The distribution of these various respiratory pigments within the animal kingdom is briefly summarized in Table 10. Obviously it cannot be said that the tetrapyrrole ring is essential for oxygen transport; the hemocyanin of the squid carries out this function with considerable efficiency. Here is seen an example where, although a given type of compound performs a function so well as to suggest a unique degree of fitness, the analogous function is carried out elsewhere by other com-

[6] See Blum, H. F., *Photodynamic Action and Diseases Caused by Light*, New York (1964) Hafner.

pounds that are not homologous. No doubt many other instances of this kind could be found.

The tetrapyrrole structure appears also in the prosthetic group of a number of enzymes that are widely distributed among plants and animals. A major pathway for aerobic respiration involves an enzyme system that includes several iron-containing pyrrole compounds, the *cytochromes*. In some instances the tetrapyrrole structure is probably identical with that found in chlorocruorins. The ability to couple loosely with O_2 is associated in the cytochromes, as is that of oxygen transport by hemoglobin or chlorocruorin, with the readily reversible oxidation of the iron which is combined within the tetrapyrrole structure of the heme. Here is suggested a close relationship between analogy and homology. The cytochrome system is widely distributed among living organisms. Probably it is requisite for all aerobic metabolism, in which it serves together with dehydrogenase enzyme

Table 9.

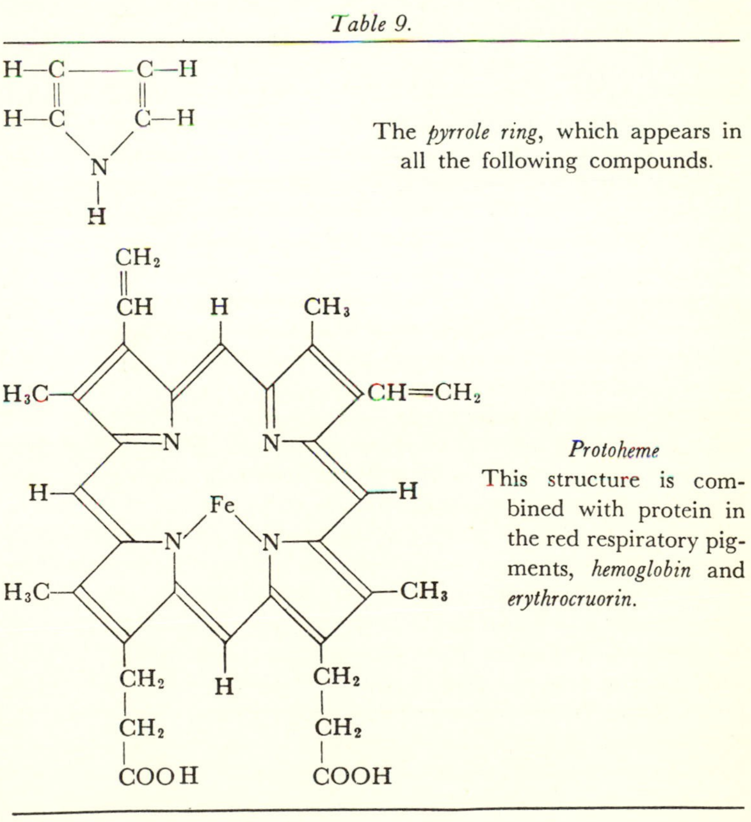

The *pyrrole ring*, which appears in all the following compounds.

Protoheme

This structure is combined with protein in the red respiratory pigments, *hemoglobin* and *erythrocruorin*.

Table 9. (*Continued*)

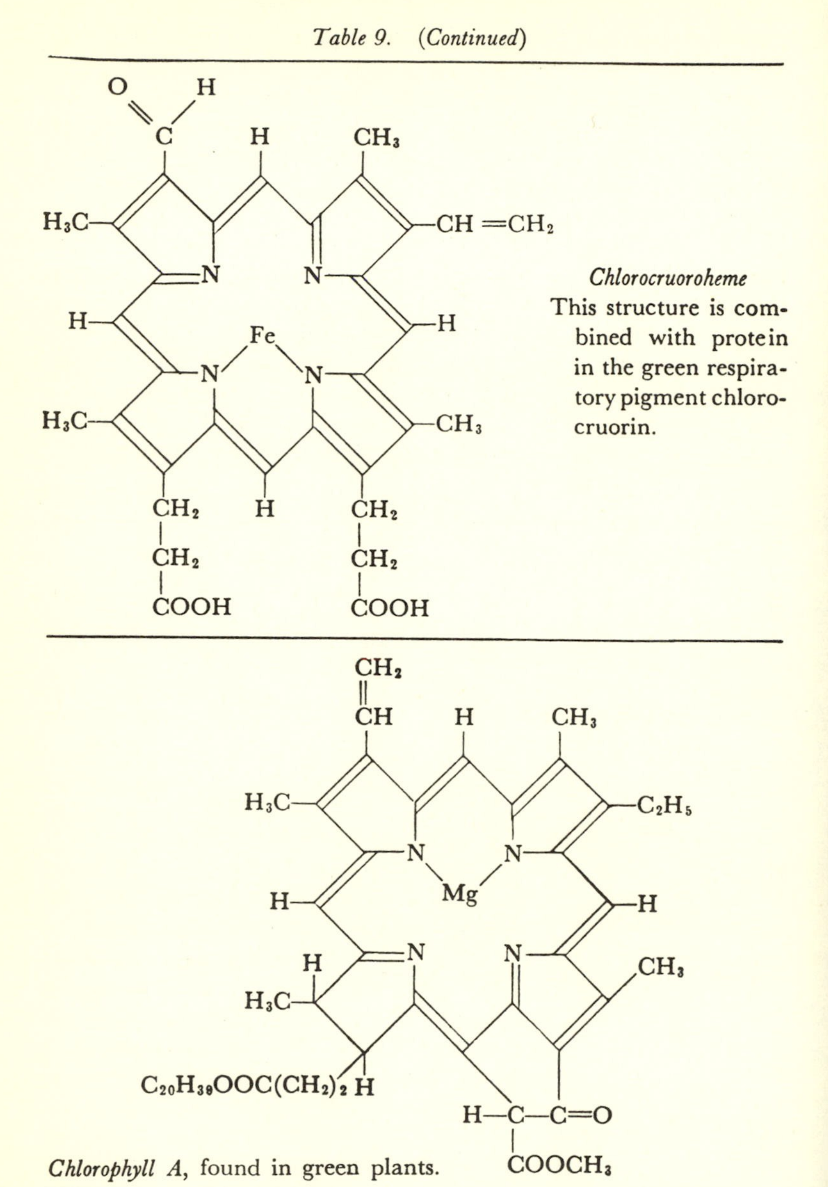

Chlorocruoroheme
This structure is combined with protein in the green respiratory pigment chlorocruorin.

Chlorophyll A, found in green plants.

Table 9. (*Continued*)

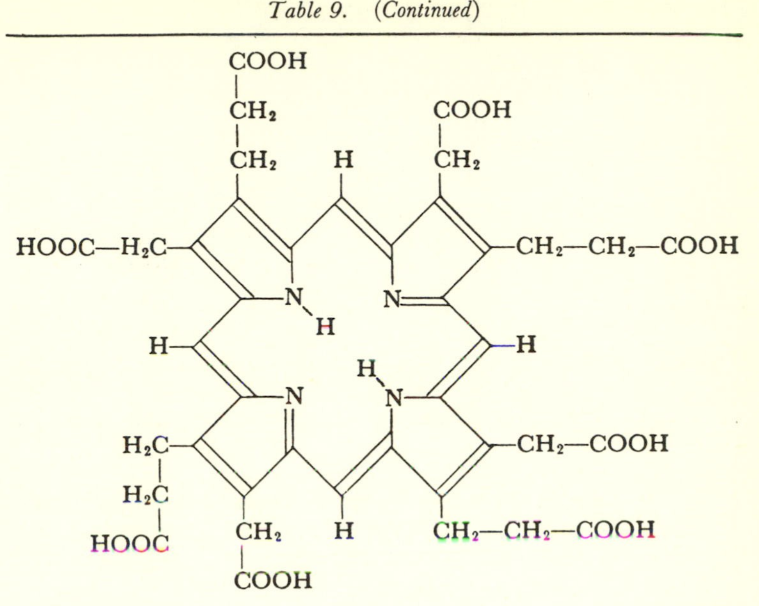

Uroporphyrin I. This compound is found in pathological conditions in man; and in the feathers of some birds.

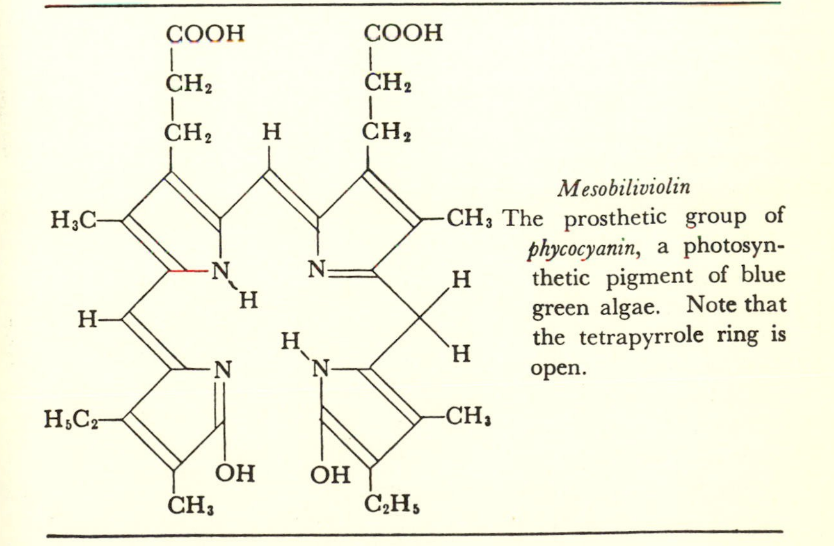

Mesobiliviolin
The prosthetic group of *phycocyanin*, a photosynthetic pigment of blue green algae. Note that the tetrapyrrole ring is open.

Table 9. (*Continued*)

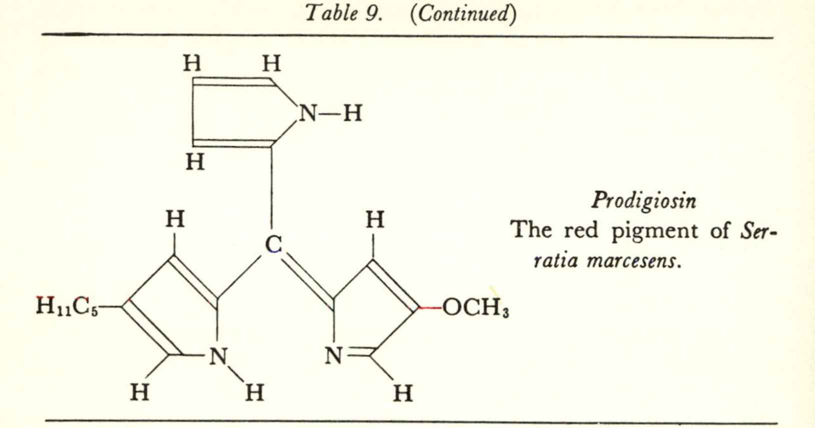

Prodigiosin
The red pigment of *Serratia marcesens.*

Table 10. Distribution of Respiratory Pigments Among Animal Groups.

Phylum	Hemoglobin[1] "invertebrate" (erythrocruorin)	Hemoglobin[1] "vertebrate"	Chlorocruorin	Hemerythrin	Hemocyanin
Protozoa		+			
Porifera					
Coelenterates					
Nematodes	+				
Annelides	+		+	+	
Arthropods	+				+
Molluscs	+				+
Echinoderms	+				
Chordates					
Protochordates					
Cyclostomes	+				
Fishes		+			
Amphibians		+			
Reptiles		+			
Birds		+			
Mammals		+			

[1] Hemoglobin has also been found in the root nodules of leguminous plants.

systems in making it possible for O_2 to act as a hydrogen acceptor. Its ubiquitous role is only emphasized by the absence of cytochromes from anaerobic organisms, and the fact that certain blood parasites which have not the ability to synthesize the tetrapyrrole ring must have it supplied to them from the outside. For aerobic life *Hemophilus influenzae* must have heme compounds supplied to it, but this same organism can carry on anaerobic existence in the absence of substances containing the tetrapyrrole ring.

Whereas it seems probable that the cytochrome system has been performing a fundamental role in aerobic metabolism since remote time, the function of O_2 transfer to the tissues, as fulfilled by the hemoglobins and chlorocruorins, did not come into prominence until very much later when circulatory systems were developed by animal organisms. To what extent did natural selection have to do with introducing the tetrapyrrole ring into this role? The employment of these compounds in oxygen transport is probably as definite a characteristic of the vertebrate animal as one might choose. The prosthetic group of the hemoglobin is always the same, protoheme, and it is this that confers the ability to carry oxygen in loose combination; but each vertebrate species has its own type of hemoglobin, the differences being referable to differences in the globin part of the molecule. Differences in the properties of the hemoglobins may confer better adaptation to a given kind of life. For example, the effect of CO_2 on the oxygen-carrying capacity of the blood, known as the Bohr effect, differs from one species to another, and may be linked with adaptation to environmental conditions. It is probable that natural selection has been largely responsible for such adaptation. But the tetrapyrrole pattern is the same throughout, even to the number of protoheme units per globin unit, with the exception of the lowest group, the cyclostomes, whose hemoglobin resembles that found in lower forms. We may believe that the hemoglobin pattern was adopted early in the evolution of the vertebrates, and that it has not been subject to alteration or substitution since. This continuance of the same basic pattern while other evolutionary changes of great magnitude were taking place should give the appearance of direction in evolution; it might be described as an orthogenetic trend.

In the case of the invertebrates the picture is quite different, however. The various invertebrate hemoglobins appear in widely separated species, where there seems to be no evidence of genetic relationship. The haphazard distribution of these pigments among the major groups of the animal kingdom is suggested in Table 10. A more detailed table would only emphasize the sporadic nature of their occurrence; for example, only one species of gastropod mollusc (*Planorbis*) possesses hemoglobin (erythrocruorin). Chlorocruorin, which contains another type of tetrapyrrole ring, appears only among the annelid worms. It seems as though the ever-present tetrapyrrole structure had been utilized over and over again to build pigments for transporting O_2 to the tissues, or possibly in some of these instances for storing a supply of this metabolite. Natural selection has without doubt been the dominant immediate factor in this adaptation to environmental needs; but at each step natural selection has worked with the pattern already at hand, it has not created a new one. Apparently natural selection has also utilized other unrelated structures to carry out the analogous function, as shown in the case of hemerythrin and hemocyanin, the distribution of which within the animal kingdom is also shown in Table 10. Here again, examination of the details of the distribution makes it appear even more haphazard then the table suggests. Orthogenetic trends would be difficult to find in this instance. Rather, there is evidence that a certain few patterns have been used quite unsystematically for analogous functions. It may be supposed that the possession of a compound of this kind capable of forming a loose combination with O_2 has in each instance had adaptive value in fitting the organism to its environment—that through mutation, probably in a series of steps, the formation of such substances has come about, and has been preserved through natural selection.[7] But the essential fitness of the tetrapyrolle compounds for this role is related to their chemical structure, and it is difficult to see just how natural selection would have functioned in the original occurrence of these compounds. To understand how this particular structural pattern came into being we might have to trace back through a good deal of evolutionary history, probably back to the time of the origin of life, and possibly—for a complete understanding—beyond into the period of evolution of the nonliving world.

The property of loose combination with O_2 will not account for all the roles played by the tetrapyrrole structure. At least one enzyme that contains it, lactic acid dehydrogenase, catalyzes a reaction that does not seem at all analogous with O_2 transfer or transport. Again, it is difficult to fit the magnesium-containing tetrapyrrole structure, chlorophyll, into any analogy with the heme compounds. If we knew more about the intimate mechanism of photosynthesis, we might be able to do so, but at present no close relationship of mechanism is obvious. The ability of the tetrapyrrole structure to enter into photochemical reactions might be invoked here, but this general property

[7] The presence of hemoglobin in Paramecia and particularly in the root nodules of leguminous plants would seem to have little adaptive significance, but is apparently an example of a fortuitous outcropping of this pattern.

is shared by any number of other biochemical substances. The recent convincing demonstration that pigments other than chlorophyll may act as light absorbers for photosynthesis is of interest in this regard. The chromoproteins *phycocyanin* and *phycoerythrin* have as their respective prosthetic groups the open tetrapyrrole compounds *mesobiliviolin* (see Table 9) and *mesobilierythrin*. These structures are not strictly homologous with the chlorophylls, which have closed tetrapyrrole rings. Phycocyanin serves as light absorber for photosynthesis in at least some blue-green algae, a group which includes the earliest plants of which we have record, and the only ones known from the pre-Cambrian. Phycoerytherin serves in the same capacity among red algae. The carotenoid pigment *fucoxanthin*, which bears no relationship to the tetrapyrrole ring structures, may also play an analogous role in some algae. *Bacteriochlorophyll*, which is truly homologous with the chlorophyll of green plants, is concerned in photosynthetic processes that are not strictly analogous. The evolutionary relationships might be difficult to trace in these cases.

We come now to sporadic instances in which tetrapyrrole ring compounds, *porphyrins* (see Table 9) seem to have no particular function other than as coloring agents. Birds eggs are often colored by *oöporphyrin*. *Uroporphyrins*, which are found in human excreta in certain pathological conditions, give color to the wings of a few species of birds. *Conchoporphyrin* has been isolated from the shell of the pearl mussel (*Pteria radiata* Lamarck), the only place this compound is known to occur. The pyrolle ring structure is found in other arrangements than the closed four-membered ring. The open tetrapyrrole ring compound, *biliverdin*, is commonly known as a pigment occurring in the bile of mammals, but it is also found distributed among invertebrates as well as vertebrates. It appears as the blue coloring matter, *oöcyan*, in some birds eggs. *Prodigiosin* is a tripyrrole compound (see Table 9) produced by *Serratia marcescens*.[8] Its red color has been connected with the mediaeval miracle of the Bleeding Host. These instances all seem like more or less fortuitous "outcroppings" of a basic pattern, rather than important adaptive characteristics preserved by natural selection.[9] Moreover, we know that such a basic pattern is almost universally distributed in such fundamental substances as the cytochromes. The existence of such a widely distributed pattern associated with definite chemical properties would seem to offer the possibility

[8] Formerly called *Bacillus prodigiosis*.

[9] One may ask whether the fact that the tetrapyrrole ring compounds are often highly colored is not their most significant aspect in these cases. Other types of compounds not absorbing within the spectrum of human vision might escape attention, and concern as to whether or not they have adaptive value.

for serial evolution along adaptive lines, as in the case of the hemoglobins of the vertebrates. At other times, however, this pattern might manifest itself in apparently nonadaptive aspects of living organisms, such as the above instances of coloring pigments seem to suggest.

Perhaps a still more basic pattern is that of the nucleic acids, which are present in all living systems without exception, and from which certain other substances of great biological importance are derivable. The nucleic acids are built of smaller units known as nucleotides, each containing a nitrogenous base, a pentose sugar, and phosphoric acid. This type of structure has been encountered in the adenylic acid system, which was discussed earlier at some length because of its importance in energy metabolism. The structure of a member of this system, ATP, appears in Table 11. A similar compound of general importance in biological oxidation is Coenzyme I. Both these compounds contain a particular structure, the purine ring. A related structure, the pyrimidine ring, is found in thiamine (vitamine B_1), and its phosphate Cocarboxylase. The nucleotide pattern is certainly a most basic one,

Table 11.

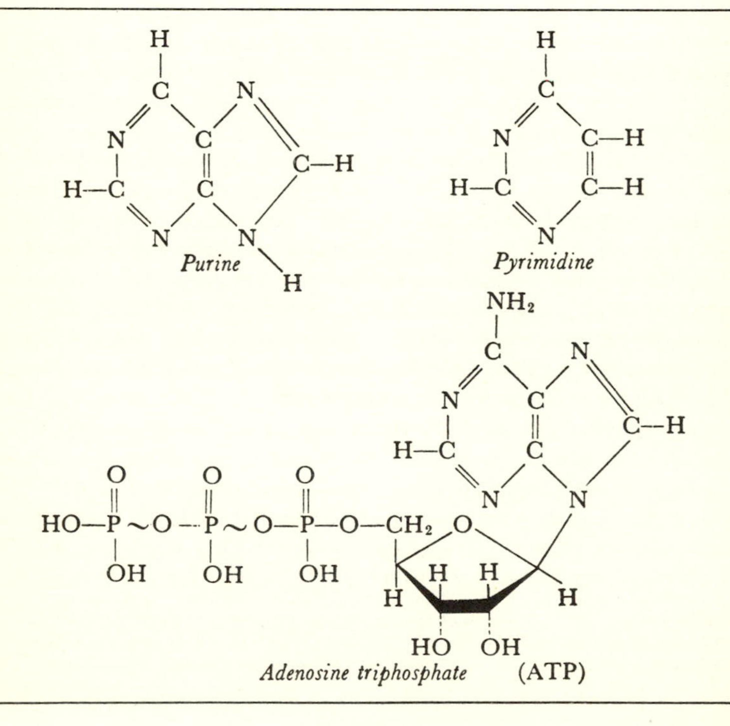

Table 11. (Continued)

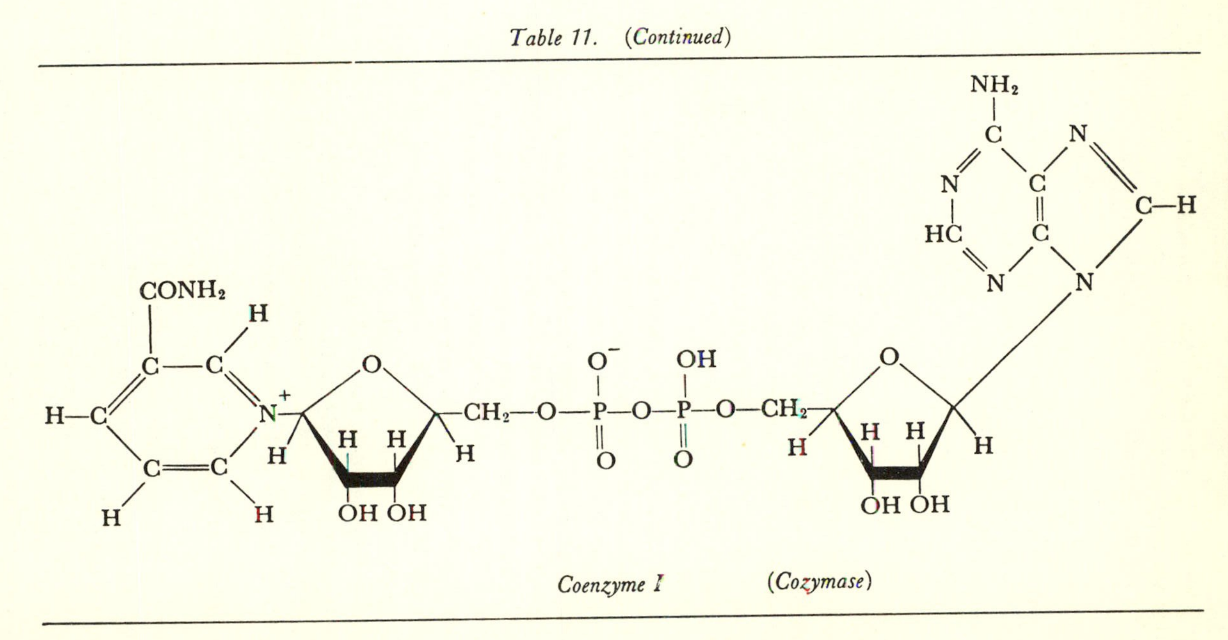

Coenzyme I (*Cozymase*)

Table 11. (*Continued*)

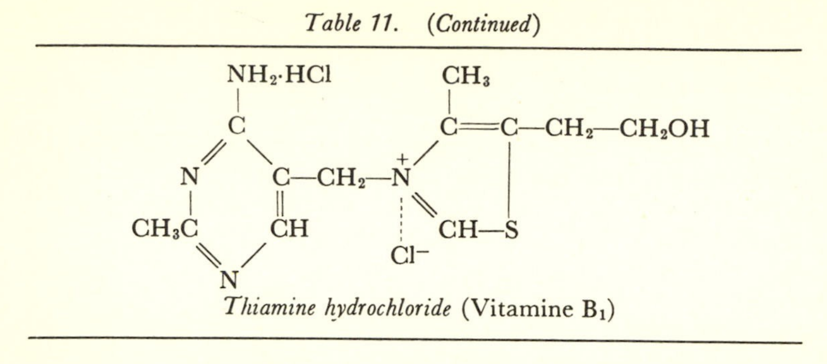

Thiamine hydrochloride (Vitamine B_1)

which it would be reasonable to think was present in the "primordial life stuff." Indeed, it was suggested in the last chapter that the nucleotide ATP may have been an essential constituent of the primitive free-energy mobilizing system, and that it possibly derived from non-living matter. If so, it could have provided the basis for the first reproducible pattern, essential to living systems from their very beginnings, and synthesized by all of them ever since.

The discussion of other patterns, perhaps less basic although widely distributed nevertheless, might be continued at length.[10] An intriguing example, the steroid structure found in vitamin D, in various sex hormones of the mammal, and also distributed among lower forms of animals and in plants, might deserve particular mention. But without understanding a great deal more about the workings of living machines and the actual manner in which they synthesize these structures, we cannot know what the recurrence of any pattern really means. Nevertheless, evidence such as that described above suggests that once living organisms acquire a basic reproducible pattern they are likely to repeat it. The pattern may be expected to appear from time to time in adaptive roles where it will be preserved by natural selection, thus giving the appearance of direction in evolution.

Mutation and Natural Selection

There has been frequent reference in preceding pages to natural selection as a factor leading to fitness of the organism to its environment. Our use of the term may at times have seemed tinged with animism, for it readily lends itself to metaphor. It is now time to consider its meaning more carefully. Natural selection describes a process by which those organisms possessing heritable characters that

[10] See, for example, Florkin, M., *Introduction à la biochimie générale*, Paris (1946) Masson; and *A Molecular Approach to Phylogeny*, (1966) Elsevier.

make them better fitted to their environment are permitted to increase more effectively than their less well-adapted fellows. Thus, the better adapted forms tend to crowd out and supplant the less fit. It would seem that such a factor must alone account for an apparent direction in evolution, direction toward better fitness of the organism to the environment. It is, of course, implicit in the idea of evolution by natural selection that inherited variations occur; these are often referred to as the "handles" for natural selection.[11] The distinction between the gene and the inherited character, which is the expression of the action of the gene, has been schematized diagrammatically in Figure 17. It is upon the inherited characters—or more properly the integrated sum of these characters which makes up the phenotype—that natural selection impinges. As the direction of the arrows in the diagram indicates, natural selection has no direct effect on the genes themselves. That is to say, although natural selection can change the proportion of genotypes in the population by suppressing the survival of certain ones, there is no direct effect of the environment on the genes themselves in the sense that mutation of a gene is guided in the direction of greater fitness.[12] In analyzing the situation, then, the probability of mutation and the relationship of the effects of that mutation to the environment may be treated as separate entities.

We may suppose that when living systems first emerged on this earth, the molecules carrying the information for their replication were more directly exposed to the environment than is the case with modern organisms—we may even imagine that at one time genotype and phenotype were indistinguishable. Natural selection would then have been more direct. An alteration of molecular pattern which conferred better fit with the environment could have been quickly expressed in the population of organisms, and rapidly produced diverging species. Under these conditions evolution might also have been relatively reversible. The separation between phenotype and genotype would seem to have increased progressively, with the emergence of bisexuality, the disposal of genes in organized chromosome systems, the pairing of genes, the dominance of one pair in determining a given phenotypic character, etc. Correspondingly, the distribution of geno-

[11] There are, of course, many variations that are not inherited, but these are of no importance in biological evolution, where the individual is of little importance as such. Cultural evolution, in which the human individual would seem of the greatest importance, is a very different matter, which should not be confused with the present topic.

[12] Direct effect of the environment upon inheritance is the essence of "Lamarckian" ideas. It was implicit in Darwin's "pangenesis", a concept he developed long after publication of *The Origin of Species* (see *The Descent of Man*). This is today a matter of historical interest only; modern Darwinism stems from Darwin's earlier ideas, and has no place for pangenesis.

types in the population would have become progressively more complicated and the establishment of mutation in them subject to more restrictions. The distribution of modern genotypes follows well-established rules which form the basis of *population genetics*, including the Mendelian and Hardy-Weinberg laws. Studies of genotype distributions, both in the laboratory and in the field, leave no doubt that evolution by mutation and natural selection occur today, although it may come about more slowly and with more restrictions than when patterns of organisms and of populations were simpler. The achievements in the field of population genetics cannot be gone into here, but it may be pointed out that with such complicated population systems, the chance that evolution will return for any distance over the pathway it has traveled becomes decreasingly likely. Evolution would seem to have become progressively less reversible, and the apparent directions it has taken less likely to be deviated from. Irreversibility of evolutionary processes will be taken up again in Chapter XIII.

XII · SOME IMPLICATIONS

"For dysteleogy is hardly less obvious in nature than teleology. . . ."—
LAWRENCE J. HENDERSON

WE HAVE traveled a considerable distance along a path that may at times have seemed quite haphazard. Parts of the terrain we have passed through have been explored in some detail, but others have hardly been touched. For our purpose has not been to exhaust our subject—which seems inexhaustible—but to gain insight and orientation regarding the retracing of the course of time's arrow into the remote past where no fossil record exists, back to the very dawn of life and beyond. No doubt some of the trails we followed were false, and the interpretations in error too. But this should only make it increasingly clear that a knowledge of things that happened at the time of the origin of life, and even long before, would be essential to a real understanding of the factors that have shaped the course of organic evolution. Our path now leads us inevitably into the borderland between biology and philosophy, and we can hardly break off our journey without at least a cursory look at this region. We can hardly hope to find answers to metaphysical questions, but we may learn things helpful in framing them.

Regarding Time, Organization, and Complexity

Difficulties encountered in the study of living organisms are often attributed to their "complexity" and "organization." What do these terms mean? I suppose the first means that such systems are composed of many and diverse facets; and the second, that these facets have some orderly mutual arrangement. Thermodynamically, order or organization is measured in terms of the number of possible arrangements that the parts of a system may enjoy. If the system can be described in terms of only a few arrangements of its microscopic parts, we say that it is very orderly. If these parts enjoy great freedom in their arrangement, so that they can only be described in terms of many possible arrangements, the system is said to be highly random or disorderly. The degree of randomness determines the entropy. A crystal at absolute zero temperature may have only one possible order for its parts; the entropy is then zero. A gas at ordinary temperatures displays high randomness in the distribution of the atoms or molecules of which it is composed; it has high entropy. The second law of thermodynamics says that left to itself any *isolated system* will tend toward an increase in its entropy or randomness. Yet we see living systems developing and maintaining what appears to be high complexity and organization, out of what seem relatively randon surroundings. Does this mean that they do not obey the second law of thermodynamics, which we take for granted applies to all nonliving things? It has been pointed out repeatedly in the foregoing chapter that such a question only arises if we fail to grasp what is implied in the term "isolated system" when used in a thermodynamic sense, that is, a system which is isolated from exchange of energy with its surroundings. But since this is not an uncommon error—particularly among those who seek philosophical implications—it may not be amiss to illustrate with a few examples by way of summary.

Let us imagine as our system a saturated solution of cane sugar in water, contained within a vessel. This system is allowed to cool somewhat, with the result that sugar crystals begin to form. The solution itself is in a relatively random state, the molecules being able to move about freely, and to occupy a great many positions with respect to each other. When a crystal begins to form, this particular part of the system becomes highly organized, in that the molecules of the crystal occupy quite exact positions relative to each other. The number of arrangements they can enjoy is relatively small, so the crystal has lower entropy than the adjacent solution. We witness, then, in the formation of the crystal, a decrease in entropy within a small part of the system, and should expect an increase in entropy somewhere else. To find where this increase occurs, it may be necessary to enlarge the system beyond the limits of the vessel containing the solution, to include that part of the surroundings into which heat has flowed as the vessel cooled. Only then do we deal with an isolated system, and we find that although a certain small part of that system comprised by the crystal has decreased in entropy, there has been an increase in entropy and randomness within the system as a whole.

Let us take another example: a volume of sterile aqueous solution containing appropriate inorganic ions and a certain amount of sugar. This system would seem to have relatively high entropy. We now introduce a single heterotrophic microorganism capable of growth under these conditions. The microorganism grows and divides until soon the solution contains a great number of microorganisms all like the first one. Surely the system now displays a higher degree of organization than before. But actually there has been a degradation of energy in the system as a whole. The sugar has undergone chemical reaction, its chemical potential having been released in part as the free energy which was utilized by the microorganisms in reproducing themselves. Again there has been an increase in total randomness in the system, an increase in entropy, although a certain part of the system upon which our attention has been focussed now seems more highly organized than it was before.

Let us take yet a third example: a volume of water containing CO_2 and certain inorganic ions only. This is an even simpler system than we started with last time, since it contains no substance of high chemical potential such as sugar. We introduce a single *photosynthetic* microorganism, and place the system in the sunlight. Soon the microorganism has reproduced itself many times, and we have a culture containing perhaps millions of these organisms. We must admit that this system, consisting of the solution and the microorganisms has really increased in organization; it has really decreased in entropy. But is it after all an isolated system? Where must we look for the increase in entropy that should compensate that represented by this increase in organization? We have to go all the way to the sun, which we must include in our isolated system. For the source of the energy utilized in reproducing the microorganisms stems from nuclear reactions in the sun, which have entailed increase in randomness. In all these three instances, the latter of which corresponds very closely to the case of living organisms as a whole, we see increase in total organization only when we view a restricted part of the universe. If we enlarge our system enough to treat it as a thermodynamically isolated one, we find sooner or later an increase in randomness. When we think of the high organization of living organisms, we need to remember that we deal with a small part of a much greater whole.

In the latter two examples there occurs an increase in total organized living matter within the system we have chosen to study, but does this really illustrate the problem of organization? In each of these cases we have deliberately introduced the pattern of that organization into the system when we seeded it with a microorganism, and as a result there has been a quantitative increase in the organization of the system as a whole. But there is no evidence of qualitative increase in organization—no introduction of new pattern—so these examples tell us nothing about how organization originates. This seems to introduce another aspect of the problem, but before going further along this line, let us examine the problem of complexity, which seems closely entangled with that of organization.

If all things tend continually toward a condition of greater randomness, which would seem to represent a tendency toward increasing uniformity, how can complexity increase even in small parts of a system? Certainly if the tendency toward greater randomness flowed along smoothly in all things, at a uniform rate, the resulting course of events would be a most monotonous one. The earth is the interesting place it now is, because this is not true. The rate of flow of events in the direction pointed by time's arrow is not constant or uniform; some events move more rapidly than others in the direction of greater randomness. Thus, although the increase in randomness may be taken as a measure of direction in time, it cannot be taken as a measure of the rate of passage of time, because the rate is different for each process examined. With different processes going at different rates, greater apparent complexity may result during finite periods of time and within restricted areas of the system as a whole. We might expect that regions in which there was local decrease in entropy—such as those described above—might also be regions in which complexity could develop, that is, increase in organization and in complexity might be expected to go hand in hand.

But now we come to see that to explain the organization and complexity of living things we would need to follow through the course of many asynchronous processes, in which the integrated whole we are interested in studying changes from one instant to the next. Viewed in this way it seems unlikely that, by examining them only in the fleeting instant of time which we as individual men may witness, we shall ever be able clearly to understand the real essence of the patterns of organization and complexity which living systems present to us. Only by going back over the course of evolution and examining each pattern in terms of the one that preceded it could we expect to comprehend completely the meaning of what we now see. Obviously we can do this in only the roughest sort of extrapolation, although we may hope that increasing understanding will come with increasing perspective. Truly, to understand living systems we must take into account their fourth dimension, and this of course is what we try to do when we study their evolution.

Regarding Fitness

There seems to be an orderly interrelationship of things in the world we know, that may be spoken of as fitness. Lawrence J. Henderson stressed the fitness of the environment for living systems. These systems themselves display a high degree of fitness to that environment. At this point the relationship seems a mutual one; the two things inseparable. Nevertheless the attempt to divide fitness into component parts for analysis may help us to understand these interrelationships. For this purpose I find it useful to distinguish three general aspects of fitness. The first has to do with the basic nature of the physical elements in the universe as a whole. The second has to do specifically with the proportions, combinations, and arrangements of these elements in our own earth, where the environment impinges directly upon the living systems. The third aspect is the fitting of living systems themselves to this environment. Each of these aspects has its history; all merge in the integrated evolution of fitness.

Consideration of the first aspect leads us back to the primordial constitution of matter and energy, and the origin of the elements. There is a general consistency in the make-up of the explored parts of the universe which leads to the idea that the elements themselves have evolved from some more generalized material, conceivably from an undifferentiated mass of the particles—protons, electrons, neutrons, etc.—of which they now are composed. It has been suggested that the evolution of the elements from such a mass was initiated at the moment of beginning expansion of the universe. Whatever their history, and however established, the constitution of the elements seems to be characteristic of the universe as a whole. We expect hydrogen to be hydrogen, and iron to be iron, wherever in the universe they may be, granted the conditions are such that they can exist as elements. The properties and general proportions of the elements establish the first order of the fitness we see about us.

The second aspect concerns more specifically the proportions and combinations of these elements which determine the character of our planet, earth. Our solar system may be essentially like any number of other solar systems in the universe, but the relationships of our planet are different from any of the other planets in our system. Many aspects of fitness are intimately tied up with these relationships, and with the intimate individual details of the evolution of the earth. When we think of the fitness of hydrogen we think of something that derives from the essential nature of the universe. When we think of the fitness of water, we think of something that is restricted to a very small part of the universe; for the existence of that substance on the earth may involve "evolutionary accidents" of only local importance. We might properly call this a second order of fitness.

The third aspect involves the fitting of the living system to its environment. This is an active process in that the system concerned changes to fit its environment—not *in order to* fit its environment, but as a result of mutation and natural selection. Implicit in this aspect is the existence of organized systems which reproduce themselves with great faithfulness, but with occasional individuals undergoing minor variations that are themselves faithfully reproduced. Those variant types that are best fitted to the environment are favored, and come to dominate and eventually supplant less fitted ones. Such factors find no parallel in the other two aspects of fitness, since there is no mechanism by which reproducible components of nonliving systems form persistent reproducible variants independently of the direct action of the environment. But variants extending in all directions are not possible in living systems either, there being restrictions placed by the nature of the materials of which the living organisms are composed, and the way in which they are combined and arranged. These restrictions stem from the other aspects of fitness that have been mentioned. The present composition of living systems is the ultimate result of their past history, extending back to the moment of their origin from nonliving material itself. Before that time mutation and natural selection could not exist, since they depend upon properties which are characteristic of living systems only, so the derivation of the basic materials must have involved other factors. The nature of these basic materials must have placed restrictions on the channels within which living systems might reproduce and might vary, and hence gave some degree of direction to all subsequent evolution. Thus, although the details of fitness of the organism to the environment seem to result from the "active fitting" that results from mutation and natural selection, it must be recognized that restriction to certain channels has played an inexorable role in the general shaping of this fitness.

I like to compare evolution to the weaving of a great tapestry. The strong unyielding warp of this tapestry is formed by the essential nature of elementary nonliving matter, and the way in which this matter has been brought together in the evolution of our planet. In building this warp the second law of thermodynamics has played a predominant role. The multicolored woof which forms the detail of the tapestry I like to think of as having been woven onto the warp principally by mutation and natural selection. While the warp establishes the dimensions and supports the whole, it is the woof which most intrigues the aesthetic sense of the student of organic evolution, showing as it does the beauty and variety of fitness of organisms to their environment. But should we pay so little attention to the warp, which is after all a basic part of the whole structure? Perhaps the analogy would be more complete if something were introduced that is occasionally seen in textiles—the active participation of the warp in the pattern itself. Only then, I think, does one grasp the full significance of the analogy.

There is, perhaps, something besides aesthetic satisfaction to be got from an examination of the warp of evolution. The judgement of accuracy of fit between organism and environment is to a certain extent a subjective one. Preconceptions regarding the mechanism by which the fit is accomplished may influence one's estimates of the degree of fit between organism and environment, and the idea that mutation is entirely random and without restriction may lead to an overemphasis of the fit that can be accomplished by natural selection. This may be, from a philosophical point of view, as misleading as is the Lamarckian notion of variation specifically induced to attain such fit. Recognition that there are restrictions placed upon the direction of mutation of organisms by their very physical make-up may introduce an appropriate skepticism in this regard. Purely physical factors impose restrictions on the direction of organic evolution, and must have imposed such restrictions from the very beginning of life, since that beginning was inextricably tangled up with the physical nature of the then existing earth. And so too is the very nature of living systems today inextricably entangled with their origin and evolution, and with the origin and evolution of the earth, and of the universe.

Regarding Teleology

The course of a few mediaeval centuries witnessed a striking evolution of cathedral architecture, particularly in what is now northern France. Starting with the round-arched Romanesque style, which could only build stone vaults over small squarish areas, developing through the Norman, was reached the Gothic, which flung its stone vaults high into the air to the joy of men's hearts and the greater glory of Mary the Virgin. Looking back today, with the perspective of several intervening centuries, we may trace an orderly evolution from the one form to the other, in which short-lived divergences from the main channel do not confuse us. May we find some analogy here to the evolution of living organisms? In both instances each new step was dependent upon those that had preceded, but there seems very little evidence, even in the architectural analogy, that the final achievement was foreseen in advance. The solution of the problem of stone vaulting of irregularly shaped areas seems now a rather simple geometric trick, yet it opened up aesthetic and structural possibilities that could hardly have been anticipated at the time. Indeed, it made possible the achievement of what today looks as though it might have been a goal from the beginning: the vaulting with stone of areas large enough to accommodate great crowds of worshipers. But did the architects who began the movement away from the old Romanesque in the direction of what would many years later be the new Gothic really have this goal in sight? Certainly one can hardly believe that they foresaw how the thing would be accomplished. It seems possible that when the foundations of many of the older cathedrals were laid, there was every expectation of simply roofing them with wood, as had been the practice in large structures for many centuries before. Only long after the deaths of the original planners did the introduction of the pointed gothic arch make possible the vaulting of the wide naves of the great cathedrals. And so, the evidence of a logical evolution of architectural design, while it may lead to the semblance of purposeful direction toward some ultimate goal, is far from proof that there really was an overall purpose projecting successive generations of men's minds toward a goal clearly seen from the beginning.

Similarly, we may see in organic evolution semblances of purpose and of progress toward a goal, but the architectural analogy must give us cause for skepticism. Order we find, and fitness, and evidence that order and fitness have been arrived at by an evolutionary process, which we make some shift to explain; but purposes and goals are other matters. When we accept the idea that living organisms as well as nonliving evolve in a general direction pointed by time's arrow, we see some reason for an appearance of purposiveness. That is, evolution goes only one way in time, and at the point to which it has led us we see greater fitness than has characterized the past. As scientists we can hardly be satisfied with explaining this increase in fitness in terms of extra-physical purposive factors, or goals foreseen in advance. Yet again and again such factors and goals are proposed as necessary for the understanding of evolution; and not too infrequently teleological implications creep into biological reasoning, and even, perhaps, into the design of experiments.

But what is teleology? Regardless of its derivations and of accepted definitions, the term seems clearly to have different connotations for different minds. To Lawrence J. Henderson teleology involved the development of existing order and fitness through an evolutionary process, participated in by natural selection but also involving other strictly physical factors. His recognition of the physical factors seems his great contribution to evolutionary thought. He considered the problem of teleology in *The Fitness of the Environment*, and more particularly in his *Order of Nature;* but although his critics have sometimes attributed to him ideas of purposiveness, careful reading of these works fails to reveal them. Many a biologist—physiologists seem particularly prone to do so—has asked himself, or his students, "What is the purpose (or function) of such and such an organ?". If pressed he will no doubt have explained that what he really meant by purpose was the fitness of the organ with respect to the organism as a whole, probably justifying himself by evolutionary reasoning. Such thinking has often, no doubt, led to fruitful discovery, for reasoning in terms of purpose we may discover fitness and so be lead to understanding of underlying mechanisms; but it has also, perhaps, led us frequently into error. Is this teleology? If so, I am afraid many of us are teleologists, although I do not like the term applied to myself.

The kind of teleology which assumes purposes and goals as realities is of another sort, and is perhaps more deserving of the name. Since it is based upon extra-physical concepts this kind of teleology lies beyond the confines of science. And there, too, belong attempts to explain evolution as guided by purpose toward a goal of perfection, such as the concept of "telefinalisme" described in the later works of the late Lecomte DuNouy, and those of his numerous finalistic predecessors.

Regarding Uniqueness

How unique in the universe is the dynamic entity we call life? This question has been pondered again and again by scholars from all disciplines, with varied and conflicting outcome. One suspects that individual temperament, as well as the current thought of the period, may often have directed judgement in this regard. There are probably those who prefer to occupy a unique place in the universe, enjoying the prestige of such a position. Others may find comforting companionship in a universe peopled with an infinite number of souls like their own. Perhaps there are still other motivating ideas. Certainly there are not many facts upon which to formulate an opinion, and these require a good deal of interpretation. Among scientists within recent years, biologists, who might be expected to be directly interested, seem to have stated their opinions less often than physicists and astronomers.[1] Even among the latter there has been wide divergence of opinion.

Whether one admits that there are many solar systems like ours may depend upon the accepted hypothesis regarding the origin of such systems. The collision theory made our solar system a unique place, because a collision such as was required to initiate its formation would have been a very rare event. Other hypotheses would make the development of solar systems a common evolutionary process, or at least set no limit upon the number that might be created. But granting that there may be many solar systems similar to ours scattered throughout the universe, the question remains as to how many of these contain planets like enough to the earth to permit the initiation and evolution of life. When one considers the unique and "accidental" place of the earth within the solar system—and particularly those aspects which seem most limiting as regards the kind of life we know—he may be

[1] A notable exception is Arthur Russel Wallace, often credited with the joint discovery with Charles Darwin of the principle of natural selection. Toward the end of his life Wallace wrote *Man's Place in the Universe*, New York (1910) McClure Phillips and Co., which in terms of then existing knowledge was an able treatise sustaining the thesis that the earth is the unique abode of life.

cautious in assuming that life exists in many other places. If he considers the complexity of living systems and the combination of physical limitations and apparent accident that have characterized the course of organic evolution, he may be still more critical of the idea that life exists elsewhere in the universe.

The reader may question whether such a point of view gives sufficient attention to the possibility that life might have developed elsewhere along some other lines. He may be thinking particularly about Henderson's suggestion that life could have developed in an ammonia environment instead of a water environment. The only approach to an answer to this question seems to be to ask the questioner to imagine what such manner of life would be like. He should, perhaps, keep in mind such phases of the fitness of our own particular environment as have been discussed in Chapter VI, and such accidents as it seems necessary to invoke to explain, for example, the left-handedness of the proteins. Perhaps if one is going to try to imagine such living systems he had better redefine the term, life. For life is what we know, what we see in a sort of instantaneous photograph taken at a given point in the evolution of that life, plus what inferences we make by tracing its history back a few billion years and extrapolating beyond. It seems reasonable to believe, when we look at the matter in this way, that if something paralleling what we know as life has evolved elsewhere in the universe, it probably has taken a quite different form. And so life such as we know may be a very unique thing after all, perhaps a species of some inclusive genus, but nevertheless a quite distinct species.

And perhaps for this reason alone, this life-stuff is something to be cherished as our proper heritage. To be guarded from destruction by, say, the activity of man, a species of living system that has risen to power and dominance through the development of a certain special property, intelligence. Such a development—vastly exceeding that of any other species—has apparently given this particular system the ability to determine its own destiny to a certain extent. Yet at the moment there are all too many signs that man lacks the ability to exercise the control over his own activities that may be necessary for survival.

Since the above paragraph was written, the pendulum has swung even farther toward the idea that the universe is full of populated worlds. Newer cosmological theories permit estimates of the number of planets that reach enormous figures, and it is frequently assumed that many of these planets are inhabited by living things. The imagination immediately pictures those living things as resembling the ones we know here on earth; and, perhaps without being aware, we envision man-like creatures distributed throughout the universe. If we think, however, of the delicate balance of conditions our earth enjoys, and to what extent chance has entered repeatedly into biological evolution, it seems that the probability of evolving a series of living organisms closely resembling those we know on earth may be a relatively small number. This becomes poignantly evident when we think of all the chance events concerned in the evolution of the human brain—which occurred only once on our planet. For close parallelism of biological evolution among the planets Time's Arrow would have had to play a much more directly deterministic role than now seems likely. So for the present I think I will let my statement stand (1954).

The pendulum has now (1968) swung still farther. The notion of a universe peopled with little folk seems to appeal to post-sputnik imaginations, and the cracking of the genetic code has pushed hard in the direction of deterministic explanation. Yet evolutionists may harbor doubts and have cogent questions to ask; the "probabilistic" approach in the next chapter is concerned with some of these.

XIII · ORDER, NEGENTROPY, AND EVOLUTION

◆◆

"*. . . that dark miracle of chance which makes new magic in a dusty world.*"—THOMAS WOLFE

◆◆

To study evolution in detail we must, perforce, regard it as a distinct entity, taking it out of the context of a great deal of the world in which it occurs. This is intuitive, methodologically necessary, and safe enough within limited scope. But when we try to extend our view to include broader relationships, or to think more rigorously in terms of thermodynamics and probability, we encounter difficulties. Although we know much more about the coding of genetic information and quantitative aspects of natural selection among modern populations than we did when this book was first written, there still remains much to be learned about the basic nature of evolutionary processes. And even with the further advances that are to be anticipated in coming years, there may, in the end, be unresolvable uncertainties. Recognizing these, we may continue to progress in our studies. Failure to recognize the uncertainties may lead us along false pathways to faulty conclusions.

One of the uncertainties enters into the drawing up of a thermodynamic balance sheet. The total of living material forms a very thin layer on the surface of the earth, the *biosphere*, which is continually being degraded and reconstituted out of the same mass of chemical constituents. The work involved in this reconstitution and in those evolutionary changes that go on in the biosphere is supported by photosynthesis, the process through which a small fraction of the energy of sunlight is captured; thus the biosphere system is dependent upon energy exchange in the sun-earth system of which it is a part. The energy of sunlight received by the earth is ultimately reradiated to space; consequently the earth's temperature remains nearly constant. But the outgoing quanta are smaller in size and greater in number than the incoming, this representing a continuing increase in entropy of the sun-earth system. Since any increase in order within the biosphere must be very small compared to the increase of entropy in the sun-earth system there is no reason to think that evolution controverts the second law of thermodynamics, even though it may appear to do so if viewed as a thing apart. Quantitative summing of the entropy changes, however, presents difficulties. For this we should require an index of units of order, which could fit within the dimensional framework customarily used in thermodynamics.

Order and Negentropy

Let us review our concept of order. If we picture a number of things that can be distributed in a variety of ways within a given space, we may say that the fewer places these things can occupy the more orderly the system; the more places the things can occupy the more disorder. "A place for everything and everything in its place," should express the quintessence of orderliness, perhaps only to be found in a perfect crystal at zero degrees K.

In thermodynamics it is customary to conceptualize order in terms of the number of arrangements of microscopic properties in a well defined system—we cannot actually determine this number but we may imagine that we can do so. The greater the number of possible arrangements the greater should be the disorder or randomness; the smaller the number the greater the order. The second law of thermodynamics predicts that a system left to itself will, in the course of time, go toward greater disorder. Following (III-10) we may represent such a change in a system made up of N molecules by,

$$\Delta S = N\mathbf{k} \ln \frac{W_2}{W_1}, \qquad \text{(XIII-1)}$$

where ΔS is the entropy change, $\mathbf{k}$ is the Boltzmann constant, and W is the number of arrangements of microscopic properties.

This does not imply that local parts of the system may not for a time increase in order. This may take place when a crystal forms from a cooling super-saturated solution. Under these conditions the molecules that form the crystal find places in the crystal lattice, whereas they were previously distributed randomly in the solution; this represents an increase in order in the crystal itself. When molecules go from solution to crystal, the number of arrangements of their microscopic properties is reduced. Let us express this for a system which is a part of one such as described in (XIII-1) by

$$-\Delta s = n\mathbf{k} \ln \frac{w_1}{w_2}, \qquad \text{(XIII-2)}$$

where Δs, n and w are all smaller than their counterparts in (XIII-1); $w_2 > w_1$.

Changing the sign (XIII-2) becomes

$$\Delta s = n\mathbf{k} \ln \frac{w_2}{w_1}. \qquad \text{(XIII-3)}$$

This represents the local increase in entropy if the crystal dissolves and the molecules return to the solution, or the decrease in entropy that must occur elsewhere in the system to just compensate the change in the reverse direction indicated in (XIII-2) when the crystal forms.

It is to be noted that ΔS and Δs measure changes in *relative* order, e.g., the change in W relative to W_1. The term *negentropy* is used in information theory to connote increase in relative order in a very different kind of system from that just discussed. In principle, a digital computer works by choosing one of two possible answers to a question posed to it; each successive question being predicated upon the answer to the preceding one. Thus each answer may be regarded as contributing a unit of relative order to the arrangement of knowledge or *information.* Using the number of questions answered as a measure,

$$\text{negentropy} = \log_2 \frac{b_2}{b_1}, \qquad \text{(XIII-4)}$$

where $\log_2$ indicates the logarithm to the base 2, and b is the number of questions answered. The binary unit expressed here is referred to as a *bit* of information.

Although (XIII-4) resembles the expressions for entropy change, it is clear that negentropy cannot be equated to $-\Delta s$ as the equation is written. The term has been used in more general sense, however; and as we shall see, the idea of successive choices exemplified by computer operation may be useful in considering natural selection and other evolutionary processes—the term *choice* is employed here and throughout this discussion without any connotation of extra-physical interference, that is, in the same way natural selection is used. So before discussing the possibility of giving negentropy thermodynamic expression let us essay a more general view in which computer operation may be regarded as a special case.

Facet and Pattern; a Model

When we have to deal with such imponderables as have just been mentioned it becomes clumsy and not particularly enlightening to think in terms of arrangements of microscopic properties, although it can hardly be denied that these are involved. We may reduce our difficulty perhaps by thinking more abstractly, viewing such examples of increase in order as the fitting of *facets* into *patterns.* For example, the facets may correspond to choices made by the computer in achieving a pattern of information, to selections of mutations which ultimately determine the pattern of a phenotype, or to innovations that are received into and help to build up a pattern of human culture. In any case, the pattern is not, like a jig-saw puzzle, already complete and needing only to be put together; it evolves progressively. Each choice of a facet depends upon the preceding *sequence of choices* by which the pattern was built up. The facets too may not exist as such until they become a part of the pattern; if this seems obscure remember that in the formation of a crystal, the arrangement of microscopic properties that is characteristic of the molecules when they are a part of the lattice is not the same as it was for the molecules when they were still in solution. We must, of course, be careful not to take the analogy too literally in a physical sense, but to regard both facet and pattern as abstractions only to be compared very loosely with such things as molecules and crystals.

The incorporation of facets into a pattern restricts the places they may occupy, representing an increase in order. And as the pattern grows it may be expected that the number of facets that can be added will increase proportionally, just as the number of molecules that can be added to the crystal increases proportionally as the crystal grows. Let us express the supposed relationship by:

$$-\Delta s = q \ln \frac{f_2}{f_1}, \qquad \text{(XIII-5)}$$

where f is the number of facets and q is a proportionality constant; $f_2 > f_1$.

It is clear that (XIII-5) cannot be equated to (XIII-2) since q cannot be put into the same coordinate framework of dimensions as $\mathcal{N}$ and **k**, nor f treated as a reciprocal of w. We may imagine that this can be done, however; and by similar imagining, and changing the base of logarithms, we may put (XIII-4) in terms of (XIII-5). We may thus apply the term negentropy to changes in relative order of various kinds; but in doing so we must not think that we can quantitatively relate any change of this kind to any other, nor to negative entropy changes in thermodynamically definable systems. We should not forget that we have imagined our way out of a difficulty, lest we come to more rigid conclusions than are warranted. Whether, for example, we see evolution as entailing either increase or decrease in order depends upon the dimensions of the system in which we choose to view it, and about these dimensions we remain, most of the time at least, very uncertain.

Figure 21 may be of help in relating the facet-pattern model to some aspects of evolutionary processes, although it may not apply as well in some cases as in others. The shaded areas with a variety of shapes represent facets that have become aggregated to form patterns. Pattern A represents a "stage" in the formation of pattern B; the central area is to be regarded as vague in both cases. We may suppose that the clear area surrounding the patterns is loosely filled with facets, or components of facets that have not as yet been incorporated into any pattern. Let us imagine the whole system to be continuously shaken by some sort of machine, so that the unattached facets move

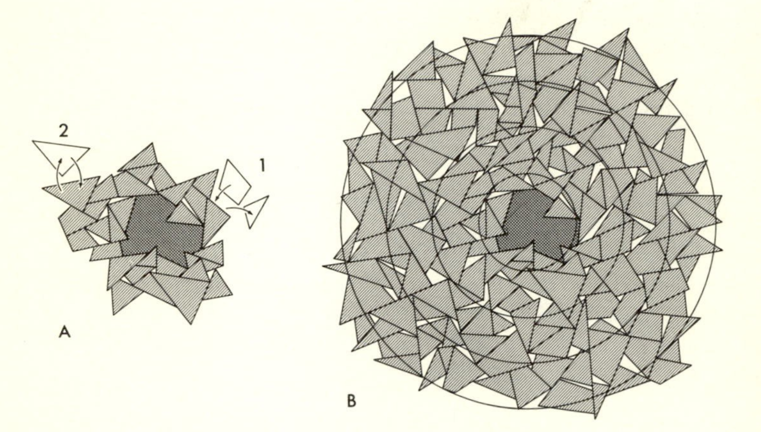

FIGURE 21. Schematic diagram to represent the formation of patterns by the addition of facets (see text).

about in random fashion. Now and then an unattached facet comes close enough to an established pattern to become attached, provided it is compatible in shape and that facet and pattern are mutually sticky enough at the place of attachment. The chance that a facet will become a part of the pattern thus depends not only upon the facet's coming close enough as the result of random movement, but upon shape and stickiness, which are specific properties of facet and pattern. We may think of such an event as a choice between the particular facet that has been accepted into the pattern and another facet that might have come close enough but did not have as appropriate shape and stickiness; this is indicated at 1, where one facet is about to be accepted, another rejected—analogy to computer operation is noted. Such choices may be regarded as based upon both chance and specific properties. Choice of an alternative facet, as indicated at 1, would have resulted in a different pattern, but one which would somewhat resemble B, since it started from A.

The attachment of a facet to the pattern may be relatively reversible, as indicated at 2 where the arrows suggest movement of a facet to and from its position in the pattern. But the chance of reversal becomes less as the pattern enlarges and the facets tend more and more to restrict each other's movement. The interstices between facets suggest the possibility for internal readjustment within the pattern; should such occur the compactness of the pattern and the fixation of the facets might be expected to increase, tending also toward irreversibility.

If we think of patterns as being replicated, as living systems are, the possibility of choice could result in the separation of numerous species of patterns. Since these will have been the same up to a given point—for example, A in the figure—they must resemble each other to some extent. The result should be an array of species of pattern, all basically related. The model permits the possibility that patterns might start off differently at the very "beginning," but if we assume competition between patterns, say for the supply of facets, some patterns might be favored and others suppressed. Choice would again be involved in this competition. Should we regard only the surviving patterns, the suppressed ones being lost sight of, we would see an array of species of pattern which would appear to increase in order. Up to a certain point analogy may be drawn to evolution by natural selection, but let us not press this too far.

Guided by the concentric circles in the diagram, the likelihood of adding facets is seen to increase proportionally as the pattern grows. This is in accord with (XIII-5); and within the conditions outlined above the increase in relative order represented by growth of pattern in our model could be assigned a value for $-\Delta s$, and the term negentropy applied.

Implications and Analogies

Energy and Negentropy. In the above model the increase in negentropy is made possible by the shaking which induces the random movements of the facets; so in drawing up a thermodynamic balance sheet the source of energy supply for the shaking machine would have to be included. In similar treatment of computers, the source of the electrical energy for the machines must be accounted for; virtually all of this comes ultimately from sunlight, whether stored in coal and petroleum in past ages or as water power (we might have to include today a very small amount coming from nuclear reactors or tides). Also included should be the metabolism of the men who program the computers, which also comes from sunlight. Thus a thermodynamic balance sheet for computer operation should be related to energy exchange in the sun-earth system, as are all the evolutionary changes taking place in the biosphere. In no case do we find controversion of the second law of thermodynamics if we enlarge our view enough.

Although the increase in negentropy is always dependent in one way or another upon the expenditure of energy, the two things are not measurable in the same terms and cannot be equated. For example, how would one set about relating number of bits to energy used by the computer? Yet sometimes the terms "entropy" and "negentropy" are confused with "energy" and this may lead to very wrong conclusions.

Negentropy and Probability. The relationship between entropy and probability is shown in (III-10) and (III-12), from which we may write for a system of $\mathcal{N}$ molecules

$$\Delta S = \mathcal{N}\mathbf{k} \ln \frac{W_2}{W_1} = \mathcal{N}\mathbf{k} \ln \frac{P_2}{P_1} \qquad \text{(XIII-6)}$$

where P is the probability of change in order within the system.

Or for a system such as the sun-earth which cannot be put into the same coordinate framework

$$\Delta S = Q \ln \frac{P_2}{P_1}, \qquad \text{(XIII-7)}$$

where Q describes the dimensions of the system.

Correspondingly we might, following (XIII-5), write for an included part of such a system in which order increases:

$$-\Delta s = q \ln \frac{f_2}{f_1} = q \ln \frac{p_1}{p_2} \qquad \text{(XIII-8)}$$

where p is the probability of change in this included system.

Thus we see a decrease in probability within the system described by (XIII-8) if we view it as only a part of the system described by (XIII-7) and do not try to put the two into the same dimensional framework.

But if we disregard the shaking machine or other source of energy, focusing our attention only on the pattern itself, does not the likelihood of adding facets increase as the pattern grows? Does not the likelihood of further increase in knowledge increase with each question that is answered by the computer? Do not the chances of cultural innovation increase as the pattern of culture expands? And would not increase in negentropy be expected to occur in any system containing parts which, because of their specific properties, may aggregate into orderly patterns, and where some means—always, of course, involving energy expenditure—exists for bringing them into juxtaposition? But may we not find a semantic dilemma here, which results from restriction of perspective by focusing upon a system out of context with a larger one of which it is an obligatory part?

Surely it is useful to think in terms of the likelihood of change with reference only to the included system, for example, of evolution by natural selection within the biosphere without reference to the sun-earth system. It might be proper, then to introduce a new term consistent with our use of "negentropy," "facets," and "pattern," or even "information" and "bits." I would suggest the word *expectability*, defined by

$$-\Delta s = \rho \ln \frac{\epsilon_2}{\epsilon_1} \qquad \text{(XIII-9)}$$

where ϵ is the expectability and ρ is a proportionality constant; $\epsilon_2 > \epsilon_1$.

We may thus think of the expectability of evolutionary change without being concerned with its probability in the larger system in which it takes place.

Facet and Meaning. Negentropy, as the term is used in information theory, is concerned with number of units of information (XIII-4), but not with the amount of meaningful content to be associated with these units. Similarly, the number of facets accumulated into a pattern in our diagramatic model does not take into account shape and stickiness. Yet it is on the basis of shape and stickiness that the choice of facets is made. Similarly the choice of answers given by the computer depends upon the meaningful content contributed by the answers to preceding questions, and natural selection depends upon properties of the phenotype and the environment. In all these cases it may be difficult to distinguish sharply between properties and their arrangement, but it is useful for purposes of analysis to assume that we can do so.

It would seem then that increase in order is a result rather than a cause of evolutionary change. In this case it would be wrong to think that either entropy or negentropy per se impels evolutionary events in a given direction. There is no basis for predicating a "driving force for evolution"—an expression currently common although even as metaphor it seems out of harmony with the parlance of science.

Certainly, meaningful content (which we may think of as a property) plays a major role in the choice of events in evolution in a real world. And we may expect that the addition of facets to pattern, in whatever context these terms may be used, is dependent upon meaningful content, which for brevity may be called *meaning.* And so in any evolutionary process involving choice we may expect that meaning and the number of facets in a pattern will increase more or less proportionally.

As commonly used (outside of Information Theory) the term information would seem to include meaning, rather than just number. This usage is perhaps justified—I shall follow it below—so long as we do not come to think that number measures something it cannot. In practice we do not, of course, measure true meaning, but instead use some index appropriate to the data in hand, which may be called *attributed meaning.*

Chemical Evolution. In Chapter X, I have discussed kinds of choices of pathway that might have given direction to evolution in a nonliving world, before metabolizing, replicating systems emerged. The process, based on the sequential accumulation of choices, I have termed chemical evolution, distinguishing it from evolution by natural selection which involves replication of patterns. No doubt these merge and overlap at some vague area in time where we choose to place the "origin of life"; but for purposes of analysis these two kinds of evolution should be considered separately on the basis of their widely different mechanisms.

A common analogy may be drawn, however, in terms of our facet-pattern model. In chemical evolution, choice of the chemical species that will be formed must depend both on the properties of the reactants and the milieu in which they find themselves. For example, the evolution of the polypeptide ancestors of the proteins, could take place only where there was a supply of amino-acids and conditions favorable to their combination. Either a drying-up puddle heated by the sun or a coacervate in a dilute aqueous soup might have provided such a milieu, whereas a simple dilute aqueous solution would not. Either of the first two alternatives seems to represent a relatively uncommon situation in a world with water aplenty; and a sequence of many choices of one kind or another may have been involved in achieving either situation. For example, much attention has been given to the "left-handedness" of the amino acids in proteins, since here we deal with a choice between two things that are thermodynamically indistinguishable; but must there not have been many choices made before the situation was reached where this one became a possibility?

Ingenious experiments, particularly those of Sidney Fox and co-workers, have shown that various basic components of living systems, may form from appropriate mixtures of simpler compounds under conditions compatible with our ideas of the constitution of the earth a few billion years ago. Such demonstrations give us an idea of plausible steps in the course of chemical evolution which led up to the emergence of replicating systems, but does not each one involve choices of materials and conditions? In the laboratory these choices are made by the experimenter, but in chemical evolution they would have been determined by events and situations depending upon choices occurring in a non-living world. For example, say that in some way nucleosides had been formed. These could not form nucleotides or combine into nucleic acids until phosphorus was available. And it seems unlikely that enough phosphorus became available until considerable erosion of land surfaces had occurred, which entailed, if we include the evolution of those surfaces, a long sequence of choices.

I, of course, use the term choices as defined above, where I have pointed out that these depend upon both chance and specific properties. The extent of the role played by each would have to be taken into account in estimating the true probability of chemical evolution's having taken the course it did on our earth. But any guess in this regard must admit, it would seem, that many choices were made.[1]

Evolution and Natural Selection. Analogy between evolution by natural selection and the facet-pattern model seems apparent enough if we think in terms of choices between alternative genetic patterns. At a very general level a successful mutant form may be thought of as being chosen instead of the ancestral form from which it derives. The process is not, of course, so simple and our diagrammatic representation of the facet-pattern model (Figure 21), perhaps not too applicable. A single mutation does not lead directly to a new biological species, or even a new stable variant, but represents a single unit change within a population of genes carried in a population or organisms (Chapter XI).

While the possibility of minor reversals of evolution by natural selection cannot be denied, there seems little doubt that in overall view this is an irreversible process. The chances of maneuvering back for any distance through the maze of events concerned in introducing mutations, once they have occurred, into a population of genotypes, would seem small indeed (see Chapter XI). Comparison may be made to the facet-pattern model as diagrammed in Figure 21, where one sees the possibility of limited reversal, but overall irreversibility. Replication of organisms should tend to preserve any pattern once it is established, although permitting minor alterations and even short reversals. The tendency to conserve patterns is, after all, as essential to evolution by natural selection as the possibility of mutation, and

[1] Quotations from this book were recently used out of context to support the belief that evolution could not have occurred without extra-physical interference in events involving what I have here called choices (Cook and Cook, *Science and Mormonism*, Deseret Press, 1964). I think a careful and unprejudiced reading would fail to find support herein for such mystical notions.

this would seem to be paralleled in other evolutionary processes in which order increases.

Selection on the basis of adaptation to the environment is also not simple, being determined not by a single overt characteristic of the organism, but on the basis of the relationship of all of the characteristics which constitute the phenotype to multiple aspects of the environment. Figure 22 gives a schematization of this relationship, focused

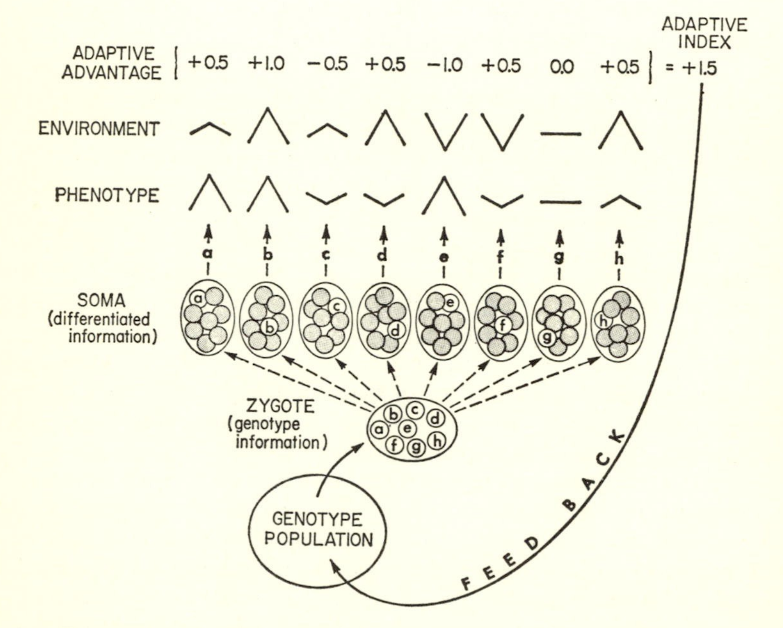

FIGURE 22. Schema to represent transfer and modification of information within the organism, and relationships with the environment and the genotype population (see text).

on one member of a population of organisms. The points where the phenotype of this organism impinge with the environment are indicated by a series of pairs of angles, one of each pair representing a characteristic of the phenotype, the other representing an aspect of the environment. The degree of fit is different for each pair, as is indicated by the corresponding plus and minus values at the top of the figure under the heading *adaptive advantage.* The algebraic sum of these values is labelled the *adaptive index.* Let us suppose that a mutation occurring by chance in the genotype of this member of the population results in a change in the adaptive advantage of a particular phenotypic character; the adaptive index will then be pushed in one direction or the other. This may cause the mutant genotype to be favored or suppressed in the population as a whole. It is on this kind of index that natural selection must be based, and it is in these terms that we may think of a feedback in organic evolution, as suggested in the figure. Phenotypic characters may be related to more than one gene—the possibility of this is suggested in the diagram as discussed below. This may complicate the three component—phenotype, genotype, environment—relationship upon which natural selection is based, so the diagram should be regarded as an oversimplification.

Let us for the moment focus attention on the impingement of phenotype and environment. The diagram indicates that some phenotypic characteristics of minor or even negative adaptive advantage may be carried along in the population if, when summed together with all the other adaptive advantages, the adaptive index has a positive value. One can only at a certain risk assume that a particular feature of the phenotype determines the fitness of the species to the environment, although this may be virtually the case in some instances. The risk is the greater because it is often infeasible to measure pertinent aspects of either the phenotype or the environment with sufficient accuracy (e.g., Blum 1961). Thus, much may have to be left to the imagination, a poor quantitative guide. Uncautious thinking in terms of single points of impingement and too exact fit between phenotype and environment may lead one unwittingly to a kind of teleological view of the best of all possible worlds.

One factor contributive to the conservation of an established biological pattern may perhaps be best understood from a consideration of the development of a complex multicellular organism, say a vertebrate animal. It might be expected that the genetic pattern (genotype) established when the egg and sperm nuclei unite to form the zygote would be distributed to all the cells that are replicated from it to form the body of the organism; we know, of course, that a similar chromosome pattern is to be found throughout. The cells of the *soma*—that is, all those not constituting the germ plasm—do not retain all the physiological and morphological characteristics of the zygote cell, but differentiate into various forms characteristic of the tissues they compose. The differentiated cells reproduce their kind in the particular tissues. For example, epidermal cells divide to form epidermal cells and muscle cells to form muscle cells, even when grown in culture outside the body. We have to assume that the original genetic pattern of the zygote has been modified in each case; whether by change in the genetic coding itself or by different interpretation of the coded message, it would be hard to say. Whatever the nature of the modification, it may occur in steps, certain groups of cells undergoing differentiation in the course of embryonic development and subsequently differentiating again into yet other types. Differentiation is essentially irreversible although dedifferentiation is thought to occur in certain cases which might be compared to the limited reversal indicated in Figure 21, at 1. Certainly there is no evidence of large scale reversal under normal conditions or that tissue cells ever return completely to the zygote pattern of the fertilized egg.

We may then distinguish between the initial *genotype information* in the zygote and the various modified versions of this found in the different tissues, which let us call *differentiated information.* Characteristics of the phenotype are determined by physiological and morphological properties of the pertinent tissues and the organs they compose, so it is in the differentiated information that the immediate plan of the phenotype is to be sought rather than in the master plan of genotype information. Thus any change in adaptation of the phenotype to the existing environment depends directly upon change in differentiated information and indirectly upon change in genotype information. These relationships are indicated in Figure 22, where the genotype is represented as composed of a set of genes labelled a, b, c, etc. Each gene is indicated in the figure as having major control over a particular phenotypic character; a common sharing of information is also indicated, since information from the genotype is passed on to the soma cells, and modified there. We may think of the modification of information as a one-way cycle, as the figure indicates. A chance modification in the genotype due to a mutation may affect differentiated information in the soma and hence alter the adaptive index. This may result in a change in distribution of genotypes in the population. But the reverse is not possible; for example, a mutation in a cell of the soma (somatic mutation), although affecting the adaptive index of the organism, would not affect the genotype of the individual in which it occurs nor the distribution of genotypes in subsequent populations.

A change in genotype information resulting from a mutation must, however, conform sufficiently with differentiated information if it is to get through and affect the adaptive index. Too great incompatibility at any point in the course of differentiation might result in death of the organism; and in the course of many non-lethal mutations resulting in successive changes in the adaptive index, the differentiated information might be so altered that some non-lethal mutations would become lethal. Differentiated information would thus serve as a kind of sieve for genotype information. This could be the basis of what Lancelot Whyte has called, "internal selection." We may suppose that alteration of the differentiated information may also occur more subtly. Since each soma cell receives a general pattern of information from the genotype, any mutation in the latter may be expected to affect the differentiated information in all the soma cells to a certain extent, even though expressing itself chiefly in one phenotypic character. Thus in the course of successive mutations within genotypes, the sieve of differentiated information would be altered, gradually affecting the distribution of genotypes in the population. In all these ways the modification of information in the soma could confer irreversibility and give direction to the course of evolution by natural selection.

Cultural evolution. The evolution of human culture—using that term to include all aspects of Man's social behavior and traditions—may be compared to the facet-pattern model, since cultural innovations can be thought of as facets which are introduced progressively into the pattern of culture as a whole. If the rate of adding facets were constant such a process would be described by

$$\ln\frac{f_2}{f_1} = r(t_2 - t_1), \qquad \text{(XIII-10)}$$

where t is time, and r is a rate constant.

A close approximation to such a growth curve is seen in Figure 23, where number of inventions is plotted against time. The data are taken from a paper entitled, "Is Invention Inevitable?" by Ogburn and Thomas, which has been sometimes cited as giving an affirmative answer to the question posed in the title. The authors list the repeated occurrence of the same (or similar) inventions—some of which may be classed as technical, some as theoretical—during five centuries ending with the beginning of our own. They conclude that re-invention has been frequent during this period. As is seen at A in the figure, however, the inventions fall closer together as time proceeds; and plotted cumulatively as at B, the data follow the drawn curve quite well, which is one predicted by (XIII-10). Plotted in this way the rate of inventions is seen to accelerate progressively, as would be the case if each invention made others likely, thus increasing the expectability of adding facets as a pattern grows (XIII-9). Interpreted in this way the data indicate that the expectability of invention increases with growth of the cultural pattern, rather than that invention is inevitable in the sense of being preordained.

The curve rises faster at some times than others, and might be compared to a curve for money at compound interest with the interest rate changing as economic conditions change. The technology that surrounds us today seems to accelerate its growth in similar fashion, as does the evolution of types of stone tools when we examine the record accumulated for a million years or more (see Blum, 1967). Numerous curves comparable to that in Figure 23B are to be found in the litera-

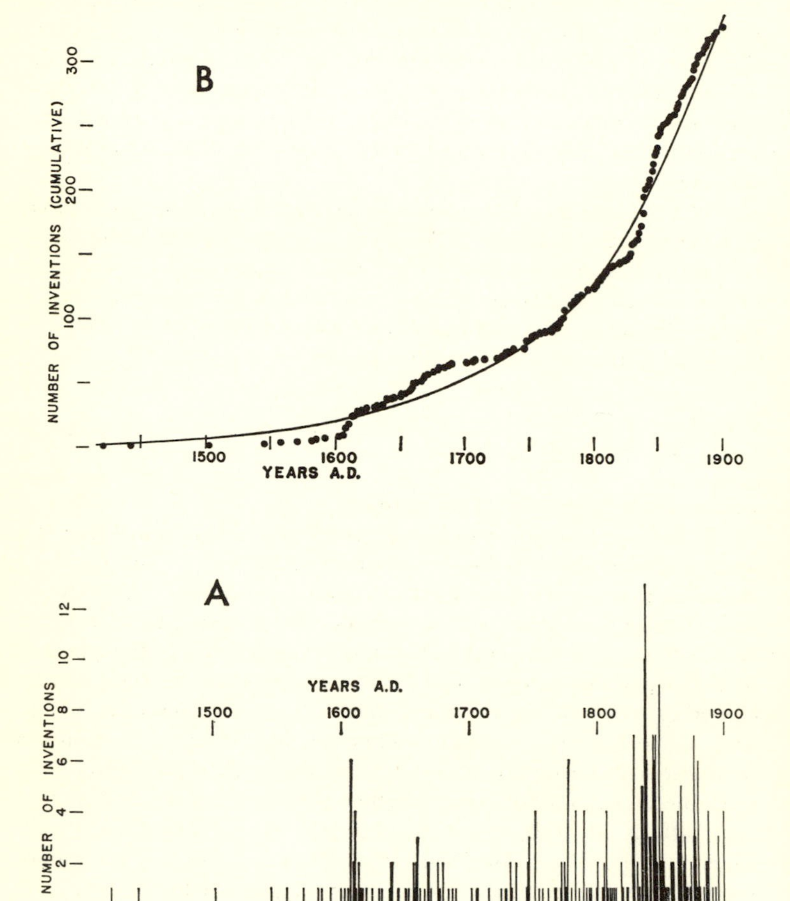

FIGURE 23. A. Number of inventions during five centuries ending at 1900. B. Plotted cumulatively. Data are from Ogburn and Thomas, *Political Sci. Quarterly* (1922) **37,** 83.

ture describing various aspects of culture. All of these are based on attributed meaning and may not be extremely accurate; but the fact that culture often grows in this way seems clear enough. As would be expected, such curves may show deviations from the simple one based on constancy of rate (XIII-10) and some curves tending to level off as though restraint were being imposed due to one factor or another (see, Blum 1967). Taking into account the many factors that could influence such curves, the situation might be better expressed in general form by

$$\ln \frac{f_2}{f_1} = \varphi(t), \qquad \text{(XIII-11)}$$

where $\varphi(t)$ is a function of time which can take a variety of forms.

Such an expression is quite non-committal, but does indicate that the addition of facets is proportional to the size of the whole pattern, just as cultural innovation seems proportional to the existing cultural pattern. This would still be true if $\varphi(t)$ assumed negative sign.

It may be asked whether relationships which might be described by (XIII-11) should not be found for evolution by natural selection, if the facet-pattern analogy holds. In general it would seem difficult to choose appropriate criteria for analysis, and it may be easier to study such relationships in cultural evolution than in biological evolution.

Analogy and differences between cultural and biological evolution. That one may draw analogy to the same model should not be taken to indicate close relationship between cultural and biological evolution, since the mechanism for transmitting information from one generation to the next is so very different in the two cases. Inherited biological information is coded accurately in long polymer molecules, even to the minor alterations resulting from mutation; it is passed on intact (subject, of course, to shuffling) from parents to their offspring. The means of transfer of cultural information is, on the other hand, much less tangible, and is likely to remain so even when we know more about the process than we do today.

Within his lifetime each human being stores information in his brain in the form of remembered images; this accumulated information I have called the *individual mnemotype*, juxtaposing the term with genotype and phenotype (1963, 1967). Information from this mnemotype is transmitted to other persons in a variety of ways—by visual, written, and other means—all of which permit more or less latitude in transmission and storage. This differs greatly from the accurate transfer of the information carried in genotypes. We may think of the cultural pattern upon which the behavior of a society is based as a kind of cross section of all the component individual mnemotypes plus information stored outside men's brains in written records etc.; this I have called the *collective mnemotype*. In the transfer and storage of information among individual members of the society and within the collective mnemotype, there is great chance for errors in copying and storage. These errors may play a role in cultural evolution comparable to that of mutations within genotypes in biological evolution, but the deviation from mnemotype patterns are much less subject to quantitative rules of transmission. Thus cultural evolution takes place with much less predictability than biological evolution; we cannot expect to describe the former in terms of Mendelian inheritance and population genetics. Cultural evolution is, of course, subject to changes in the physical environment; more directly perhaps but less compellingly than biological evolution. One would anticipate that cultural evolution, being less subject to restraining factors, could take place much more rapidly than biological evolution—a brief comparison of Figures 1 and 23 indicates that this has been true.

But there must be a tendency to conserve cultural pattern, or chaos rather than orderly evolution would result from such lack of restraint. Such a tendency may stem from the similarity of the individual mnemotypes of a given society, which are formed within the common environment provided by the collective mnemotype, where mutual exchange of information is relatively easy. Any innovation in culture, initiated by an alteration of information within an individual mnemotype, must be compatible with the traditions embodied in the collective mnemotype, if it is to be accepted into the cultural pattern. Any single innovation can produce only a small change in that pattern as a whole. As the cultural pattern enlarges the expectability of innovations increases, but only in terms of meaningful content of the pattern as it exists. This is, of course, compatible with (XIII-10), (XIII-11), and what is seen in Figure 23.

Although cultural and biological evolution have been treated separately because of the difference in their mechanisms; they are not, of course, independent. Cultural evolution which had its beginning a million years ago (speaking in round numbers) has been restricted by morphological and physiological characteristics of the animal, Man. Some of these characteristics might be traced back as far as the emergence of the vertebrates nearly half a billion years ago, or even farther. Man has continued to evolve biologically during the last million years, but cultural evolution may have played a greater role in this than vice versa. Certainly some aspects of animal behavior, generally referred to as *instincts*, have a genetic basis. This would seem also to apply to some aspects of human behavior (see Alland, 1967), but the non-genetic part dominates to such an extent that it may be difficult to sort out the genetically controlled component. Certainly physiological and morphological characteristics play a role in human behavior and these may be subject to natural selection—inherited changes in endocrine balance could, for example, have considerable influence. There is no known means by which a *specific* piece of information stored in the brain can be directly transferred to the sex cells and coded there as genetic information, as would be necessary if it were directly to affect evolution of behavior by natural selection. In accepting such a notion one would not only revert to Lamarkism or pangenesis, but would deny much of modern physiology and genetics.

The behavior of other animals shows some parallels to that of Man, although even in the closely related primates the variations may be quite wide. When close similarities are found, may not these often represent analogy rather than homology (Chapter XI)? If so, we might err greatly in assuming a common genetic basis for parallelisms in behavior or in attributing them too much to common response to the environment.

Overemphasis on genetically determined aspects of human behavior, whether real or alleged, may lend an unwarranted determinism to our thinking about Man's social behavior, in which so many non-genetic factors play a role. This kind of view seems to be on the increase, however, both inside and outside science at the present time. Once such a "fatalistic" attitude gets into the collective mnemotype it is apt to influence cultural evolution to a considerable degree, as would seem to have been the case with the false lore of "Social Darwinism."

Life on other worlds. Since the last pages of Chapter XII were written, opinion has swung still farther toward a universe peopled with life. The science of Exobiology has come into being, although without data as yet to confirm its existence. Post-sputnik researches in space have, of course, given great impulse to scientific inquiry in this direction, but may also have encouraged romantic views of the problem.

In the role of devil's advocate I recently (1966) made an estimate of the probability that life very similar to our own exists somewhere else. Estimating the number of sequential choices made in the course of our evolution to be 10^{18}, I arrived at a probability of 10^{-18} which may be contrasted with some estimates that have been given for the number of "habitable" planets. At the time I considered my estimate to be conservative, but I now find it too much so. I assigned to cultural evolution a probability of 10^{-6}, that is, I assumed that 10^{6} choices were involved; but remembering that the great libraries of the world contain more than that many books—each of which may be supposed to have contributed something to the pattern of human knowledge—I might wish to lower my estimate of probability by several orders of magnitude. The estimate of 10^{9} mutations involved in the evolution of the accepted 10^{6} species of living organisms may be over-conservative, and I should wish to raise above 10^3 the number of sequential choices prior to the origin of life. These changes would lower by several orders of magnitude my original estimate of 10^{-18} for the probability of living things similar enough to ourselves to hold converse with us.

Of course, as I pointed out at the time, any estimate based on number of choices does not take into account the meaningful content to be attached to each one; when we learn how to evaluate this I may wish to revise my estimates again. In the meantime I suppose our beliefs in this regard must continue to be shaped by the emotional content of the information in our respective individual mnemotypes.

Of Cabbages and Kings

One may suppose that Lewis Carrol's conversational Walrus would have found time to discuss not only the analogies but the differences between the evolution of cabbages and of kings. He would have found certain common ground, but sooner or later might have seen more differences than parallels. I like to think that the model described in this chapter could have helped him to decide when to stop worrying about analogies and start examining divergences of mechanism. The model ought to be most useful in just that way, pointing out that while chemical, biological and cultural evolution—and perhaps even the course of cosmic events following the "big bang"—may enjoy a certain analogy, this sooner or later breaks down. Tempting as they are, analogies may lead us sadly astray if we press them too far.

We might also fall into serious error if we try to add up all the attributes of the things that concerned the Walrus—ships and sealing wax, etc.—without being aware that we do so. It is, of course, useful practice to do this in applying information theory, but one still needs to know what he is doing. The assignment of numbers without attention to the attributes of the things they represent, may be a good basis for hypothesis, but prove a shaky one for final conclusions.

Whether living systems obey the same principles as non-living is a question that continues to trouble; the approach and motive for asking the question may be different, and consequently the answer may reflect differences in point of view and dimensions of the framework in which the problem is placed. It may be difficult to keep from losing perspective when moving from one frame of reference to another —we have seen for example, that evolution may appear either to obey or to disobey the second law of thermodynamics according to the system taken into view, and that it may not be feasible to put the one into the same kind of dimensions as the other. There is sometimes a choice of what principle one is going to apply. No one would try today to study the process of photosynthesis without the quantum concept; but, on the other hand, no one would attempt to design an optical microscope without help from the wave concept. The comparison is a trite one, but may illustrate the inadvisability of trying to measure bits of information or units of natural selection in terms of the quanta of physics. It is not that quanta are not involved, but that it becomes unprofitable to think about evolution or the transfer of information in such terms. I hope no one will suppose that these remarks indicate my backsliding into the morass of vitalism; on the contrary I find it not at all difficult to avoid, so long as one does not try to wade through.

We may do well to remember that all the concepts of science are the product of a long evolution of Man's knowledge in which there were many sequential choices based on both chance and meaning. When one looks back in this way, he sees these concepts not as constituting truth—a rather vague concept that must itself have had an evolution—but as approximations thereto, which might have been approached along some other evolutionary pathway and hence have not taken quite the same shape. If quantum theory had preceded wave theory, might not the development of the optical microscope have been delayed indefinitely, but might we not be farther advanced in photochemistry? Or just where would we be? And if Man had come to think in terms of proportionalities before he began to count his fingers, what turn might mathematics have taken? We continue to wend our faltering, evolutionary way, like the whiting and the snail not always aware that there is a porpoise just behind.

BIBLIOGRAPHY

Ahrens, L., Measuring geologic time by the strontium method. *Bull. Geol. Soc. America.* (1949) *60*, 217–266.

Alland, A., *Evolution and Human Behavior.* Garden City, N. Y. (1967) The Natural History Press.

Alpher, R. A., and Herman, R. C., Theory of the origin and relative abundance distribution of the elements, *Rev. Modern Physics* (1950) *22*, 153–212.

Anfinsen, C. B., *The Molecular Basis of Evolution.* New York (1959) John Wiley and Son.

Arnold, C. A., *An Introduction to Paleobotany.* New York (1947) McGraw-Hill.

Baldwin, E., *An Introduction to Comparative Biochemistry*, 3rd ed. Cambridge (1948) University Press.

Baldwin, E., *Dynamic Aspects of Biochemistry.* Cambridge (1948) University Press.

Ball, E. G., Energy relationships of the oxidative enzymes, *Ann. New York Acad. Sci.* (1944) *45*, 363–375.

Barron, E. S. G., Mechanism of carbohydrate metabolism. An essay on comparative biochemistry, *Advances in Enzymology* (1943) *3*, 149–189.

Beadle, G. W., Biochemical genetics, *Chem. Revs.* (1945) *37*, 15–96.

Beadle, G. W., Genes and the chemistry of the organism, *Am. Scientist* (1946) *34*, 31–53.

Beadle, G. W., Physiological aspects of genetics, *Ann. Rev. Physiol.* (1948) *10*, 17–42.

Beadle, G. W., and Tatum, E. L., Genetic control of biochemical reactions in Neurospora, *Proc. Nat. Acad. Sci.* (U.S.) (1941) *27*, 499–506.

Bernal, J. D., *The Physical Basis of Life.* London (1951) Routledge and Kegan Paul. Also in *Proc. Physical Soc.* (1949) *62*, 537.

Bernal, J. D., The origin of life, in *New Biology.* London (1954) Penguin Books, 28–40.

von Bertalanfy, L., The theory of open systems in physics and biology, *Science* (1950) *111*, 23–29.

Bethe, H. A., Energy production in stars, *Am. Scientist* (1942) *30*, 243–264.

Birch, F., Schairer, J. F., and Spicer, H. C., *Handbook of Physical Constants.* Geol. Soc. America. Special Papers No. 36 (1942).

Blum, H. F., Humanity in the Perspective of Time, *Ann. New York. Acad. Sci.* (1967) *138*, 489–503.

Blum, H. F., Evolution of the Biosphere, *Nature* (1965) *208*, 324–326.

Blum, H. F., Dimensions and Probability of Life, *Nature* (1965) *206*, 131–132.

Blum, H. F., On the Origin and Evolution of Human Culture, *Am. Scientist* (1963) *51*, 32–47.

Blum, H. F., Does the Melanin Pigment of Human Skin have Adaptive Value? *Quart. Rev. Biol.* (1961) *36*, 50–63.

Blum, H. F., On the Origin and Evolution of Living Machines, *Am. Scientist* (1961) *49*, 474–501.

Borsook, H., and Dubnoff, J. W., The biological synthesis of hippuric acid in vitro, *J. Biol. Chem.* (1940) *132*, 307–324.

Breder, C. M., A consideration of evolutionary hypotheses in reference to the origin of life, *Zoologica* (1942) 27, 131–143.

Bridgman, P. W., *The Nature of Thermodynamics.* Cambridge, Mass. (1943) Harvard University Press.

Buddington, A. F., Some petrological concepts of the interior of the earth, *Am. Mineralogist* (1943) *28*, 119–140.

Bullard, E. C., Geological time, *Mem. Proc. Manchester Lit. Phil. Soc.* (1943–5) *86*, 55–82.

Cold Spring Harbor Symposia on Quantitative Biology. (1947) *12.* Several other volumes of this series also contain pertinent material.

Daly, R. A., *Architecture of the Earth.* New York (1938) D. Appleton-Century Co.

Dauvillier, A., and Desguin, E., *La genese de la vie.* Paris (1942) Hermann et Cie.

Dingle, H., Science and cosmology, *Science* (1954) *120*, 513–521.

Dobzhansky, T., *Genetics and the Origin of Species.* 2nd ed. New York (1941) Columbia University Press.

Dole, M., The history of oxygen, *Science* (1949) *109*, 77–81.

Dubos, R. J., *The Bacterial Cell.* Cambridge, Mass., (1945) Harvard University Press.

Dunbar, C. O., *Historical Geology.* New York (1949) John Wiley and Sons.

Eddington, A. S., *The Nature of the Physical World.* New York (1929) Macmillan.

Ehrensvärd, G., *Life, Origin and Development.* Chicago (1962) University of Chicago Press.

Ehrensvärd, G., On the origin of aromatic structures in biological systems, in *Symposium sur le Metabolisme Microbien*, II. International Congress of Biochemistry. Paris (1952), 72–85.

Ephrussi, B., *Nucleo-cytoplasmic Relations in Micro-organisms.* Oxford (1953) Clarendon Press.

Eyring, H., The drift toward equilibrium, *Am. Scientist* (1944) *32*, 87–102.

Fisher, R. A., *The Genetical Theory of Natural Selection.* Oxford (1930) Clarendon Press.

Florkin, M., *L'evolution biochimique.* Liège (1944) Desoer. The English translation (*Biochemical Evolution*, New York (1949) Academic Press) has been considerably modified by the translator.

Florkin, M., Biochemical aspects of some biological concepts. I. International Congress of Biochemistry. Paris (1949). (Biochemical Society (1950) 19–31.)

Florkin, M., *A Molecular Approach to Phylogeny.* Paris (1966) Elsevier.

Florkin, M., *Aspects of the Origin of Life.* New York (1960) Pergamon Press.

Florkin, M., *Introduction a la biochimie génèrale.* 4th ed. Paris (1946) Masson.

Foster, J. W., *Chemical Activities of Fungi.* New York (1949) Academic Press.

Fox, S. W., editor, *The Origins of Prebiological Systems and their Molecular Matrices.* New York (1965) Academic Press.

Franck, J., and Loomis, W. E., *Photosynthesis in Plants.* Ames, Iowa (1949) Iowa State College Press.

Gaffron, H., Photosynthesis, photoreduction and dark reduction of carbon dioxide in certain algae, *Biol. Rev.* (1944) *19*, 1–20.

Gamov, G., *The Birth and Death of the Sun.* New York (1940) The Viking Press.

Glasstone, S., Laidler, K. J., and Erying, H., *The Theory of Rate Processes.* New York (1941) McGraw-Hill.

Goldberg, L., and Aller, L. H., *Atoms, Stars, and Nebulae.* Philadelphia (1943) Blakiston.

Goldschmidt, R., *The Material Basis of Evolution.* New Haven (1940) Yale University Press.

Goldschmidt, V. M., Geochemische Verteilungsgesetze der Elemente, *Skrifter Utgitt av det Norske Videnskaps-Akademi i Oslo, Matematick-Naturvidenskapslig Klasse* (1937) No. 4, 1–148.

Gutenberg, B., editor, *Physics of the Earth; VII, Internal Constitution of the Earth.* New York (1939) McGraw-Hill.

Haldane, J. B. S., *The Causes of Evolution.* London (1932) Longmans, Green and Co.

Haldane, J. B. S., The origin of life, in *Science and Life.* New York (1933) Harper and Bros.

Haldane, J. B. S., The origins of life, in *New Biology*. London (1954) Penguin Books, 12–27.

Hall, R. P., The trophic nature of the plant-like flagellates, *Quart. Rev. Biol.* (1939) *14*, 1–12.

Henderson, L. J., *The Fitness of the Environment*. New York (1913) Macmillan.

Henderson, L. J., *The Order of Nature*. Cambridge, Mass. (1917) Harvard University Press.

Holmes, A., The oldest dated minerals of the Rhodesian shield, *Nature* (1954) *173*, 612–616.

Horowitz, N. H., On the evolution of biochemical syntheses, *Proc. Nat. Acad. Sci.* (U.S.) (1945) *31*, 153–157.

Horowitz, N. H., Bonner, D., Mitchell, H. K., Tatum, E. L., and Beadle, G. W., Genic control of biochemical reactions in Neurospora, *Am. Naturalist* (1945) *79*, 304–317.

Hulburt, E. O., The temperature of the lower atmosphere of the earth, *Physical Review* (1931) *38*, 1876–1890.

Humphreys, W. J., *Physics of the Air*, 3rd ed. New York (1940) McGraw-Hill.

Huxley, J., *Evolution: The Modern Synthesis*. New York (1943) Harper.

Jeans, J., *The Universe Around Us*. 4th ed. New York (1945) Macmillan.

Jeffreys, H., *The Earth*. 2nd ed. New York (1929) Macmillan.

Jepsen, G. L., Mayr, E., and Simpson, G. G., editors, *Genetics, Palaentology and Evolution*. Princeton (1949) Princeton University Press.

Jones, H. S., *Life on Other Worlds*. New York (1940) Macmillan.

Jones, H. S., The age of the universe, *Proc. Roy. Inst. Gr. Br.* (1948) *34*, 210–218.

Jukes, T. H., *Molecules and Evolution*. New York (1966) Columbia University Press.

Kamen, M. D., and Barker, H. A., Inadequacies in present knowledge of the relation between photosynthesis and the O^{18} content of atmospheric oxygen, *Proc. Nat. Acad. Sci* (U.S.) (1945) *31*, 8–15.

Kavanau, J. L., Some physico-chemical aspects of life and evolution in relation to the living state, *Am. Naturalist* (1947) *81*, 161–184.

Kluyver, A. J., *The Chemical Activities of Micro-organisms*. London (1931) University of London Press.

Lanham, U. N., Oparin's hypothesis and the evolution of nucleoproteins, *Am. Naturalist* (1952) *86*, 213–218.

Larsen, H., *On the Microbiology and Biochemistry of the Photosynthetic Green Sulfur Bacteria*. Trondheim (1953) F. Bruns.

Latimer, W. M., Astrochemical problems in the formation of the earth, *Science* (1950) *112*, 101–104.

Lemberg, R., and Legge, J. W., *Hematin Compounds and Bile Pigments.* New York (1949) Interscience Publishers.

Lewis, G. N., and Randall, M., *Thermodynamics and the Free Energy of Chemical Substances.* New York (1923) McGraw-Hill.

Li, C. C., *Population Genetics.* Chicago (1955) University of Chicago Press.

Lipmann, F., Metabolic generation and utilization of phosphate bond energy, *Advances in Enzymology*, (1941) *1*, 99–162.

Lipmann, F., Acetyl phosphate, *Advances in Enzymology* (1946) *6*, 231–267.

Longwell, C. R., Knopf, A., and Flint, R. F., *Physical geology*, 3rd ed. New York (1948) John Wiley and Sons.

Lotka, A. J., *Elements of Physical Biology.* Baltimore (1925) Williams and Wilkins.

Lotka, A. J., The law of evolution as a maximal principle, *Human Biol.* (1945) *17*, 167–194.

Luria, S. E., Recent advances in bacterial genetics, *Bact. Revs.* (1947) *11*, 1–40.

LuValle, J. E., and Goddard, D. R., The mechanism of enzymatic oxidations and reductions. *Quart. Rev. Biol.* (1948) *23*, 197–228.

Lwoff, A., *L'evolution physiologique.* Paris (1943) Masson.

Madison, K. M., The organism and its origin, *Evolution* (1953) *7*, 211–227.

Mayr, E., *Animal Species and Evolution.* Cambridge, Mass. (1963) Harvard University Press.

McElroy, W. D., and Swanson, C. P., The theory of rate processes and gene mutation, *Quart. Rev. Biol.* (1951) *26*, 348–363.

Menzel, D. H., *Our Sun.* Philadelphia (1949) Blakiston.

Miller, S. L., A production of amino acids under possible primitive earth conditions, *Science* (1953) *117*, 528–529.

Muller, H. J., The gene, *Proc. Roy. Soc.* (*London*) (1947) B *134*, 1–37.

National Research Council. *Reports of the Committee on the Measurement of Geologic Time.* Published occasionally by the National Research Council, Washington.

Needham, J., Contributions of chemical physiology to the problem of reversibility in evolution, *Biol. Rev.* (1938) *13*, 225–251.

Nier, A. O., Thompson, R. W., and Murphy, B. F., The isotopic constitution of lead and the measurement of geological time, *Physical Review* (1941) *60*, 112–116.

Northrop, J. H., Kunitz, M., and Herriott, R. M., *Crystalline Enzymes.* New York (1948) Columbia University Press.

Oparin, A. I., *The Origin of Life.* New York (1938) Macmillan.

Oparin, A. I., *Origin of Life.* New York (1953) Dover Publications. (This is a reprinting of the original with a new introduction by S. Morgulis.)

Paneth, F. A., *The Origin of Meteorites.* Oxford (1940) Clarendon Press.

Parpart, A. K., editor, *Chemistry and Physiology of Growth.* Princeton (1949) Princeton University Press.

Pauling, L., *The Nature of the Chemical Bond.* Ithaca, N. Y. (1944) Cornell University Press.

Pauling, L., *General Chemistry.* San Francisco (1947) Freeman and Co.

Pauling, L., Campbell, D. H., and Pressmann, D., The nature of the forces between antigen and antibody and of the precipitation reaction, *Physiol. Rev.* (1943) *23*, 203–219.

Pauling, L., and Cory, R. B., Atomic coordinates and structure factors for two helical configurations of polypeptide chains, *Proc. Nat. Acad. Sci. (U.S.)* (1951) *37*, 235 (and several later articles in the same journal).

Pirie, N. W., On making and recognizing life, in *New Biology.* London (1954) Penguin Books, 41–53.

Pringle, J. W. S., The evolution of living matter, in *New Biology.* London (1954) Penguin Books, 54–62.

Rabinowitch, E. I., *Photosynthesis, and Related Processes*, Vol. I. New York (1945) Interscience Publishers.

Rubey, W. W., Geologic history of sea water, an attempt to state the problem, *Bull. Geol. Soc. Am.* (1951) *62*, 1111–1147.

Russell, H. N., *The Solar System and its Origin.* New York (1935) Macmillan.

Schmidt, C. L. A., editor, *The Chemistry of the Amino Acids and Proteins.* Springfield (1938) C. C. Thomas. With Addendum (1943).

Schoenheimer, R., *The Dynamic State of Body Constituents.* Cambridge, Mass. (1942) Harvard University Press.

Schrödinger, E., *What is Life?* Cambridge (1945) University Press.

Shapley, H., Cosmography: an approach to orientation, *Am. Sci.* (1954) *42*, 471–486.

Shapley, H., On the astronomical dating of the earth's crust, *Am. J. Sci.* (1945) *243A*, 508–522.

Simpson, G. G., *Tempo and Mode in Evolution.* New York (1944) Columbia University Press.

Simpson, G. G., *The Major Features of Evolution.* New York (1953) Columbia University Press.

Simpson, G. G., *The Meaning of Evolution.* New Haven (1949) Yale University Press.

Slater, J. C., *Introduction to Chemical Physics.* New York (1939) McGraw-Hill.

Smyth, H. D., *Atomic Energy for Military Purposes.* Princeton (1945) Princeton University Press.

Spiegelman, S., and Kamen, M. D., Genes and nucleoproteins in the synthesis of enzymes. *Science* (1946) *104*, 581–584.

Srb, A. M., and Horowitz, N. H., The ornithine cycle in Neurospora and its genetic control, *J. Biol. Chem.* (1944) *154*, 129–139.

Sturtevant, A. H., and Beadle, G. W., *An Introduction to Genetics.* Philadelphia (1939) Saunders.

Tatum, E. L., and Beadle, G. W., Biochemical genetics of Neurospora, *Ann. Missouri Bot. Garden* (1945) *32*, 125–129.

Ter Haar, D., Recent theories about the origin of the solar system, *Science* (1948) *107*, 405–410.

Ter Haar, D., Cosmological problems and stellar energy, *Rev. Modern Physics* (1950) *22*, 119–152.

Timoféeff-Ressovsky, N. W., *Experimentelle Mutationsforschung in der Vererbungslehre.* Dresden (1937) Steinkopff.

Umbreit, W. W., Problems of autotrophy, *Bact. Rev.* (1947) *11*, 157–166.

Urey, H. C., *The Planets, Their Origin and Development.* New Haven (1952) Yale University Press.

Urey, H. C., On the early chemical history of the earth and the origin of life, *Proc. Nat. Acad. Sci. (U.S.)* (1952) *38*, 351–363.

van Niel, C. B., The biochemistry of micro-organisms; an approach to general and comparative biochemistry, in *The cell and protoplasm.* Am. Assoc. Adv. Science, Science Press. Pub. No. 14, (1940) 106–315.

van Niel, C. B., Biochemical problems of the chemo-autotrophic bacteria, *Physiol. Rev.* (1943) *23*, 338–354.

van Niel, C. B., The comparative biochemistry of photosynthesis, *Am. Scientist* (1949) *37*, 371–383.

Vening Meinesz, F. A., Major techtonic phenomena and the hypothesis of convection currents in the earth, *Quart. J. Geol. Soc. London* (1948) *103*, 191–207.

Wald, G. Biochemical Evolution, in *Modern Trends in Physiology and Biochemistry.* New York (1952) Academic Press 337–376.

Wald, G., The origin of life, *Sci. American.* (1954) *191*, 45–53.

Watson, F. G., *Between the Planets.* Philadelphia (1941) Blakiston.

Watson, J. D., and Crick, F. H. C., The structure of DNA, *Cold Spring Harbor Symp. Quant. Biol.* (1953) *18*, 123–131.

Wheland, G. W., *The Theory of Resonance and its Application to Organic Chemistry.* New York (1944) John Wiley and Sons.

Whipple, F. L., *Earth, Moon and Stars.* Philadelphia (1941) Blakiston.

Whipple, F. L., Concentrations of interstellar medium, *Astrophys. J.* (1946) *104*, 1–11.

Whyte, L. I., *Internal Factors in Evolution.* New York (1965) George Braziller.

Wildt, R., Photochemistry of planetary atmospheres, *Astrophys. J.* (1937) *86*, 321–336.

Wildt, R., The geochemistry of the atmosphere and the constitution of the terrestrial planets, *Rev. Modern Physics* (1942) *14*, 151–159.

Woodring, W. P., Conference on biochemistry, palaeoecology, and evolution, *Proc. Nat. Acad. Sci* (*U.S.*) (1954) *40*, 219–224.

Wright, S., Statistical genetics and evolution, *Bull. Am. Math. Soc.* (1942) *48*, 223–246.

Wright, S. Genes as physiological agents, *Am. Naturalist* (1945) *79*, 289–303.

Wright, S., On the roles of directed and random changes in gene frequency in the genetics of population, *Evolution* (1948) *2*, 279–294.

Wright, S., Population structure in evolution, *Proc. Am. Philos. Soc.* (1949) *93*, 471–478.

Wrinch, D., The native protein, *Science* (1947) *106*, 73–76.

INDEX